Distribution System
Requirements for Fire Protection

AWWA MANUAL M31

Fourth Edition

**American Water Works
Association**

MANUAL OF WATER SUPPLY PRACTICES — M31, Fourth Edition

Distribution System Requirements for Fire Protection

Project Manager/Technical Editor: Martha Ripley Gray, Melissa Valentine
Production Editor: Darice Zimmermann, D&D Editorial Services
Manuals Coordinator: Beth Behner

Library of Congress Cataloging-in-Publication Data

Distribution system requirements for fire protection. -- 4th edition.
  p. cm. -- (AWWA manual ; M31)
 Includes bibliographical references and index.
 ISBN 1-58321-580-8
 1. Fire extinction--Water-supply. 2. Water--Distribution. I. American Water Works Association.

 TH9311.D565 2008
 628.9'252--dc22

                                2008006142

Printed in the United States of America
American Water Works Association
6666 West Quincy Ave.
Denver, CO 80235 Printed on recycled paper

Contents

Figures

Tables

Preface

Water distribution systems have been developed and operated for a variety of reasons. In the past, many communities recognized the need for safe, potable water that could be used for drinking and cooking. Governmental agencies or private enterprise promptly took on the responsibility of providing safe water to meet these basic human requirements.

The need for nonpotable water arose because of an increasingly sophisticated lifestyle. In many cases, the water systems established to provide potable water were augmented and enlarged to provide water for irrigation, car washing, industrial processes, and other purposes. It also became necessary to provide an adequate supply of water for fire protection and suppression. Systems that provide for fire protection needs have frequently been incorporated into the systems that provide potable water service as well as nonpotable service. As technology advances, many alternatives to this basic water supply configuration can be found. A notable alternative is the use of dual distribution systems.

The American Water Works Association has published standards for materials used in the field of fire protection for many decades. However, it was not until the early 1980s that AWWA's Committee on Fire Protection developed the first edition of this manual, which addresses the planning, design, and maintenance of distribution systems that supply water for fire protection and suppression.

This manual provides specific guidance on the design, operation, and maintenance of water distribution systems as they relate to fire protection and fire suppression activities. When the governing body of a community makes a conscious decision to use the available water supply system for fire suppression purposes, this manual should be consulted, judiciously applied, and tempered as local conditions require.

This edition of the manual updates the information, clarifies some topics, and deletes material that is no longer essential. The manual still closely parallels the first edition prepared by the Fire Protection Committee.

As was the case with the first edition, this fourth edition does not intend to describe how firefighters should use water to control fires, but rather how water utilities should design and operate their systems to maximize fire protection benefits while delivering safe, potable water to customers. The emphasis is on public water systems and not on water systems exclusively designed for fire protection. Similarly, this manual does not intend to reproduce material available elsewhere in AWWA publications such as Manual M17, *Installation, Field Testing, and Maintenance of Fire Hydrants*, or Standards C502, Dry-Barrel Fire Hydrants, and C503, Wet-Barrel Fire Hydrants.

The adequacy of a water distribution system for fire protection depends on the fire flows required. Chapter 1 describes several methods for determining required fire flows. Once fire flow requirements are determined, these are added to the other water system demand requirements. Chapter 2 discusses the impact of fire protection on distribution system design. Chapter 3 focuses on distribution storage in terms of both sizing and location. Chapter 4 discusses reliability issues arising because systems must remain in operation even when individual components are out of service. Chapter 5 describes fire suppression sprinkler systems and their effect on water requirements, particularly because they affect how water is used for fire fighting.

Appendix A lists organizations involved in fire protection and describes the organizations and their roles. Appendix B describes the relationship between the water supply system and fire insurance ratings, because one benefit of supplying needed fire flow is an improved fire insurance rating for a community.

The water utility has important partners in fire protection. The water utility must work together with local fire officials, building code officials, and others to effectively and efficiently promote fire protection. Good communication between all involved parties is essential to protect property and life.

Acknowledgments

The members of the AWWA Fire Protection Committee at the time this edition was prepared included:

Kevin Kelly, Chair

Doug Bayles, Pittsburg Tank & Tower Co. Inc., Henderson, Ky.
Bob Benish, Pittsburg/International Tank, Louisville, Ky.
Phillip Brown, American Fire Sprinkler Association, Dallas, Texas
Brad Dales, Irvine, Calif.
Chris Dubay, National Fire Protection Association, Quincy, Mass.
G.C. Garber, Pittsburg Tank & Tower Co. Inc., Henderson, Ky.
G.O. Houghton, London, Ont., Canada
Kevin Kelly, National Fire Protection Engineering, Pine Bush, N.Y.
Kenneth Kerr, Kerr Marketing Agency Inc., Brunswick, Ohio
James Lake, National Fire Sprinkler Association, Patterson, N.Y.
Melinda L. Raimann, Cleveland Division of Water, Cleveland, Ohio
Sandra Stanek, Rural Metro Fire Department, Scottsdale, Ariz.
Edward Straw, Insurance Services Office Inc., Duluth, Ga.

The AWWA Fire Protection Committee would like to thank the following individuals for their assistance in revising AWWA Manual M31.

Robert J. Bennett, Response Branch, National Fire Academy
Ray Cellemme, Apollo Valves/Conbraco Industries
Ron Coke, Master Meter
George DeJarlais, Badger Meter
Tom Gwynn, Hersey Meters
John Higdon, Apollo Valves/Conbraco Industries
Kenneth Isman, National Fire Sprinkler Association
Jack Poole, Poole Fire Protection
Jim Purzyckl, BAVCO
Paul Schwartz, Foundation for Cross-Connection Control and Hydraulic
    Research, University of Southern California
Deborah Somers, National Fire Sprinkler Association
Nicole Sprague, National Fire Sprinkler Association

This manual was developed under the guidance of the Distribution and Plant Operations Division, which included the following personnel at the time of approval:

Melinda L. Raimann, Chair

Jerry L. Anderson, CH2M Hill, Louisville, Ky.
Charly C. Angadicheril, City of Fort Worth, Fort Worth, Texas
Robert L. Gardner, Wannacomet Water Co., Nantucket, Mass.
Gerald H. Caron, Wyoming Water Treatment Plant, Holland, Mich.
Rhonda E. Harris, Pro-Ops Inc., Dallas, Texas
Gregory J. Kirmeyer, HDR Engineering, Bellevue, Wash.
George A. Kunkel Jr., Philadelphia Water Dept., Philadelphia, Pa.
Bill C. Lauer, Staff Advisor, American Water Works Association, Denver, Colo.
Melinda L. Raimann, Cleveland Division of Water, Cleveland, Ohio

Chapter 1

Fire Flow Requirements

For centuries, water has been used to extinguish fires. The inexpensiveness and availability of water are the primary factors leading to its widespread use. But, not only must water be available for fire protection, it must be available in adequate supply. As a result, the question must be asked, how much water is necessary to be considered an adequate supply for fire protection? (Milke 1980)

Most municipalities are willing to incur the higher cost for distribution system sizing because of the reduction in loss that is possible by using the water system for fire protection. Water in sufficient quantity can cool the fire; the steam can deprive the fire of oxygen, and in the case of miscible or dense fluids, water can disperse the fuel. The key question for water utilities is how large must distribution system components be to provide sufficient water for fire protection. The remainder of this manual presents methods for estimating these requirements.

IMPACT ON DISTRIBUTION SYSTEM DESIGN

The decision to provide water for fire protection means that a utility must explicitly consider fire flow requirements in sizing pipes, pumps, and storage tanks. In larger systems, fire protection has a marginal effect on sizing decisions, but in smaller systems these requirements can correspond to a significant increase in the size of many components. In general, the impact of providing water for fire protection ranges from being minimal in large components of major urban systems to being very significant in smaller distribution system pipes and small distribution systems.

The most significant impacts are installing and maintaining fire hydrants, providing adequate storage capacity, and meeting requirements for minimum pipe sizes (e.g., 6-in. [150-mm] pipes in loops and 8-in. [200-mm] dead ends) in neighborhood distribution mains when much smaller pipes would suffice for delivery of potable water only. These requirements make designing distribution systems easier for the engineer but more costly for the water utility. Other impacts include providing extra treatment capacity at plants and extra pumping capacity at pump stations.

COMMUNITY GOVERNANCE

The decision of whether or not to size distribution system components, including water lines, appurtenances, and storage facilities, for fire protection must be made by the governing body of the community. This decision is made in conjunction with the water utility if the utility is privately owned. However, there is no legal requirement that a governing body must size its water distribution system to provide fire protection. In some instances, this undertaking may be prohibitively expensive. For privately owned utilities, the distribution system would not be sized for fire protection unless such an undertaking could be shown to be commercially profitable.

The governing bodies of most communities do provide water for fire protection for a variety of reasons, including protection of the tax base from destruction by fire, preservation of jobs that would be lost in the event of a large fire, preservation of human life, and reduction of human suffering.

When a community's governing body provides fire protection, it must do so in accordance with a well-thought-out plan that will provide adequate supplies for the intended purpose. An inadequate fire protection system provides a false sense of security and is potentially more dangerous than no system at all.

FIRE FLOW REQUIREMENTS

When establishing a fire protection plan, the governing body must first select a well-documented procedure for determining the fire flow requirement. Central to providing "enough" water is a determination of how much water should be made available for any given situation. The following definition of *required fire flow* will be used in this manual: the rate of water flow, at a residual pressure of 20 psi (138 kPa) and for a specified duration, that is necessary to control a major fire in a specific structure. A complete definition of required fire flow requires a determination of both the rate of flow required and the total amount of water that must be applied to control the fire. The rate of flow and the duration of flow required may be specified by the simple equation:

$$\text{quantity} = \text{rate} \times \text{duration} \qquad \text{(Eq. 1-1)}$$

Understanding Water Use

The importance of flow rate and total quantity must be realized when attempting to understand the ways in which water is used to suppress fire. Water applied to a fire accomplishes two things. First, it removes the heat produced by the fire, thereby preventing that heat from raising the temperature of unignited material to the ignition point. Water absorbs the heat of the fire when it changes from a liquid to a gaseous state as the heat is released as steam. Second, water not converted to steam by the heat of the fire is available to cool material not yet ignited. Water also blankets unignited material, excluding the oxygen required to initiate and sustain combustion.

CALCULATING FIRE FLOW REQUIREMENTS

All fires are basically different because of random variations in the structure and contents of the burning building, exposures (configuration of adjacent structures not involved in a fire but that are to be protected to prevent the fire from spreading), weather, temperature, and length of time the fire has been burning. Consequently, numerous methods have been proposed for determining how much water is enough to suppress a fire. The following sections describe four methods for calculating fire flow requirements. These methods have been developed by the Insurance Services Office

Table 1-1 Fire flow durations

Required Fire Flow		Duration
gpm	(L/sec)	hr
2,500 or less	(158 or less)	2
3,000 to 3,500	(189 to 221)	3

Inc. (ISO),[*] Iowa State University (ISU),[†] the National Fire Academy,[‡] and the Illinois Institute of Technology Research Institute (IITRI).[§]

Responsibility for determining needed fire flows for individual structures usually rests with the local fire officials based on information provided by the owner. Rating services such as ISO may determine this flow during an evaluation for insurance purposes. For planning purposes, water departments may determine representative fire flow requirements in portions of towns for system planning, hydraulic analysis, and design.

Flow Durations

Recommended fire flow durations[¶] to be used in the four methods are given in Table 1-1. The maximum required fire flow for a single fire event is 12,000 gpm (757 L/sec).

Insurance Services Office Method

The ISO's technique for calculating required fire flow is documented in its publication *Fire Suppression Rating Schedule*. The term used in that document to describe the fire flow requirement is *needed fire flow* (NFF).

Needed fire flow (NFF). The NFF is the rate of flow considered necessary to control a major fire in a specific building for a certain duration. It is intended to assess the adequacy of a water system as one element of an insurance rating schedule. It is not intended to be a design criterion. However, it has been demonstrated that the NFF reasonably coincides with the actual flow required to suppress a fire in a real life situation.

A water supply should be capable of providing the maximum NFF within its distribution system area. In designing a new water distribution system or improvements within an existing distribution system, it is customary to provide for the NFF within the design area. However, it is very unusual for an existing water distribution system to be capable of providing every NFF within its service area.

The ISO classification of a community's water system is based on the available rates of flow at representative locations, with an NFF of 3,500 gpm (221 L/sec), or less, as determined by the application of its Fire Suppression Rating Schedule. Private and public protection at properties with larger NFFs is individually evaluated and may vary from the community's classification.

---

[*] Insurance Services Office Inc., 545 Washington Blvd., Jersey City, NJ 07310-1686.
[†] Iowa State University, Fire Extension Service, Ames, IA 50011.
[‡] US Fire Administration, 16825 S. Seton Ave., Emmitsburg, MD 21727.
[§] Illinois Institute of Technology Research Institute, 10 W. 35th St., Chicago, IL 60616.
[¶] Fire flow durations are based on the 19th edition of the National Fire Protection Association's *Fire Protection Handbook*, table 10.4.6.

Table 1-2   Values of coefficient (F) construction class

	Class	Coefficient
Class 1	Frame	1.5
Class 2	Joisted Masonry	1.0
Class 3	Noncombustible	0.8
Class 4	Construction (masonry, noncombustible)	0.8
Class 5	Modified fire resistive	0.6
Class 6	Fire resistive	0.6

Calculation. The calculation of an NFF, in gallons per minute (gpm), for a subject building, considers the construction (C_i), occupancy (O_i), exposure (X_i), and communication (P_i) factors of that building, or fire division, as outlined here.

Construction factor (C_i). That portion of the NFF attributed to the type of construction and area in square feet of the subject building is determined by the following formula*:

$$C_i = 18F\,(A_i)^{0.5} \tag{Eq. 1-2}$$

Where:

F = coefficient related to the class of construction (see Table 1-2)

A_i = effective area

Effective area (A_i). This is the total area in square-feet of the largest floor† in the building plus the following percentage of the other floors:

- for buildings of construction class 1–4, 50 percent of all other floors;

- for buildings of construction classes 5 or 6, if all vertical openings in the building have 1.0 hr or more protection, 25 percent of the area not exceeding the two largest floors‡. The doors shall be automatic or self-closing and labeled as class B fire doors (1.0 hr or more protection). In other buildings, 50 percent of the area not exceeding eight floors.§

---

* Reprinted with permission—Insurance Services Office Inc., 2006. Copyright ISO Properties Inc., 2001, 2006.

† If division walls are rated at one hour or more with labeled class B fire doors on openings, they subdivide a floor. The maximum area on any one floor used shall be the largest undivided area plus 50 percent of the second largest undivided area on that floor. NOTE: Do not include basement and subbasement areas that are vacant, that are used for building maintenance, or that are occupied by C-1 or C-2 occupancies (see Table 1-3).

‡ Reprinted with permission—Insurance Services Office Inc., 2006. Copyright ISO Properties Inc., 2001, 2006.

§ Reprinted with permission—Insurance Services Office Inc., 2006. Copyright ISO Properties Inc., 2001, 2006.

Table 1-3 Occupancy factors for selected combustibility classes

	Combustibility Class	Occupancy Factor (O_i)
C-1	Noncombustible	0.75
C-2	Limited combustible	0.85
C-3	Combustible	1.00
C-4	Free burning	1.15
C-5	Rapid burning	1.25

The maximum value of C_i is limited by the following: 8,000 gpm (505 L/sec) for construction classes 1 and 2; 6,000 gpm (378 L/sec) for construction classes 3, 4, 5, and 6; and 6,000 gpm (378 L/sec) for a one-story building of any class of construction. The minimum value of C_i is 500 gpm (32 L/sec). The calculated value of C_i should be rounded to the nearest 250 gpm (16 L/sec).

Occupancy factor (O_i). The occupancy factors, given in Table 1-3, reflect the influence of the occupancy in the subject building on the NFF. Representative lists of occupancies by combustibility class are given in Figures 1-1 and 1-2.

Exposures (X_i) and communication (P_i) factors. The exposures and communication factors reflect the influence of exposed and communicating buildings on the NFF. A value for ($X_i + P_i$) shall be developed for each side of the subject building as shown in Eq 1-3:

$$(X + P)_i = 1.0 + \sum_{i=1}^{n} (X_i + P_i) \qquad \text{maximum 1.60} \qquad \text{(Eq. 1-3)}$$

Where:

n = number of sides of subject building

The factor for X_i (exposure) depends on the construction and length–height value (length of wall in feet times height in stories) of the exposed building and the distance between facing walls of the subject building and the exposed building. This factor shall be selected from Table 1-4. When more than one exposure side exists for the subject building, apply only the largest factor X_i for that side. When there is no exposure on a side, $X_i = 0$.

The factor for P_i (communications) depends on the protection for communicating party wall openings and the length and construction of communications between fire divisions. This factor shall be selected from Table 1-5. When more than one communication type exists in any one side wall, apply only the largest factor P_i for that side. When there is no communication on a side, $P_i = 0$.

Needed fire flow. The calculation for NFF is

$$\text{NFF} = (C_i)(O_i)[1.0 + (X + P)_i] \qquad \text{(Eq. 1-4)}$$

When a wood shingle roof covering on a building or on exposed buildings can contribute to spreading fires, add 500 gpm (32 L/sec) to the NFF. The NFF shall not exceed 12,000 gpm (757 L/sec) or be less than 500 gpm (32 L/sec). The NFF shall be rounded to the nearest 250 gpm (16 L/sec), if less than 2,500 gpm (158 L/sec), and to the nearest 500 gpm, if greater than 2,500 gpm.

Classification 1

Steel or concrete products storage, unpackaged

Classification 2

Apartments	Hotels
Churches	Motels
Courthouses	Offices
Dormitories	Parking garages
Hospitals	Schools

Classification 3

Amusement park buildings, including arcades and game rooms
Automobile sales and service
Discount stores
Food and beverage—sales, service, or storage
General merchandise—sales or storage
Hardware, including electrical fixtures and supplies
Motion picture theaters
Pharmaceutical retail sales and storage
Repair or service shops
Supermarkets
Unoccupied buildings

Classification 4

Aircraft hangars, with or without servicing/repair
Auditoriums
Building material sales and storage
Freight depots, terminals
Furniture—new or secondhand
Paper and paper product sales and storage
Printing shops and allied industries
Theaters, other than motion picture
Warehouses
Wood product sales and storage

Classification 5

Chemical sales and storage
Cleaning and dyeing material sales and storage
Paint sales and storage
Plastic or plastic product sales and storage
Rag sales and storage
Upholstering shops
Waste and reclaimed material sales and storage

Figure 1-1   Typical occupancy classifications—nonmanufacturing

For one- and two-family dwellings not exceeding two stories in height, the NFF listed in Table 1-6 shall be used. For other habitable buildings not listed in Table 1-6, the NFF should be 3,500 gpm (221 L/sec) maximum. For a building protected by automatic sprinklers, the NFF is that needed for the sprinkler system, plus hose streams converted to 20 psi (138 kPa) residual pressure, with a minimum of 500 gpm. See the National Fire Protection Association Standard No. 13 (13D or 13R) for the water requirements at the base of riser for sprinklers.

Classification 2

Ceramics manufacturing
Concrete or cinder products manufacturing
Fabrication of metal products
Primary metals industries

Classification 3

Banking and confectionary
Dairy processing
Leather processing
Soft drink bottling
Tobacco processing

Classification 4

Apparel manufacturing
Breweries
Cotton gins
Food processing
Metal coating or finishing
Paper products manufacturing
Rubber products manufacturing
Woodworking industries

Classification 5

Cereal or flour mills
Chemical manufacturing
Distilleries
Fabrication of textile products, except wearing apparel
Meat or poultry processing
Plastic products manufacturing
Textile manufacturing

Figure 1-2   Typical occupancy classifications—manufacturing

Iowa State University Method

Research conducted at the Fire Extension Service at ISU resulted in the development of a rate of flow formula. This formula addresses, in some detail, both the quantity of water required to extinguish a fire and the effects of various application rates and techniques. The formula is referenced in ISU Bulletin 18. The ISU method of determining required fire flows is the oldest method discussed in this manual. It was first published in 1967.

The ISU technique referenced several Danish studies for its theoretical and statistical bases. In addition, several experiments were performed, including actual fires in small rooms. The resulting equation for computation of [needed] fire flow, G, according to this technique is:

$$G^* = \text{volume of space (in cubic feet)} \div 100$$

---

* Fire flow G in this excerpt is synonymous with needed fire flow (NFF) referenced throughout this manual.

Table 1-4   Factor for exposure (X_i)

Construction of Facing Wall of Subject Building	Distance to the Exposed Building	Length–Height* of Facing Wall of Exposed Building	Construction of Facing Wall of Exposed Building			
			Construction Classes			
			1, 3	2, 4, 5, 6	2, 4, 5, 6	2, 4, 5, 6
			Unprotected Openings	Semiprotected Openings (wire glass or outside open sprinklers)	Blank Wall	
	ft		Exposure Factor X_i			
Frame, metal, or masonry with openings	0–10	1–100	0.22	0.21	0.16	0
		101–200	0.23	0.22	0.17	0
		201–300	0.24	0.23	0.18	0
		301–400	0.25	0.24	0.19	0
		Over 400	0.25	0.25	0.20	0
	11–30	1–100	0.17	0.15	0.11	0
		101–200	0.18	0.16	0.12	0
		201–300	0.19	0.18	0.14	0
		301–400	0.20	0.19	0.15	0
		Over 400	0.20	0.19	0.15	0
	31–60	1–100	0.12	0.10	0.07	0
		101–200	0.13	0.11	0.08	0
		201–300	0.14	0.13	0.10	0
		301–400	0.15	0.14	0.11	0
		Over 400	0.15	0.15	0.12	0
	61–100	1–100	0.08	0.06	0.04	0
		101–200	0.08	0.07	0.05	0
		201–300	0.09	0.08	0.06	0
		301–400	0.10	0.09	0.07	0
		Over 400	0.10	0.10	0.08	0
Blank masonry wall		When the facing wall of the exposed building is higher than subject building, use the above information, except use only the length–height of facing wall of the exposed building above the height of the facing wall of the subject building. Buildings five stories or more in height, consider as five stories. When the height of the facing wall of the exposed building is the same or lower than the height of the facing wall of the subject building, $X_i = 0$.				

Source: Insurance Services Office Inc., 2003.

NOTE: Refer to the *Fire Suppression Rating Schedule,* published by Insurance Services Office Inc., for complete information regarding factors of exposure.

* The length–height factor is the length of the wall of the exposed building, in feet, times its height in stories.

Table 1-5 Factor for communications (P_i)*

Protection of Passageway Openings	Fire-Resistive, Noncombustible, or Slow-Burning Communications				Communications With Combustible Construction					
	Open	Enclosed			Open			Enclosed		
	Any length	10 ft or less	11 to 20 ft	21 to 50 ft†	10 ft or less	11 to 20 ft	21 to 50 ft	10 ft or less	11 to 20 ft	21 to 50 ft
Unprotected	0	‡	0.30	0.20	0.30	0.20	0.10	‡	‡	0.30
Single class A fire door at one end of passageway	0	0.20	0.10	0	0.20	0.15	0	0.30	0.20	0.10
Single class B fire door at one end of passageway	0	0.30	0.20	0.10	0.25	0.20	0.10	0.35	0.25	0.15
Single class A fire door at each end or double class A fire doors at one end of passageway	0	0	0	0	0	0	0	0	0	0
Single class B fire door at each end or double class B fire doors at one end of passageway	0	0.10	0.05	0	0	0	0	0.15	0.10	0

Source: Insurance Services Office Inc., 2003.

NOTES:
1. Refer to the *Fire Suppression Rating Schedule,* published by Insurance Offices Inc., for complete information regarding factors for communication.
2. When a party wall has communicating openings protected by a single automatic or self-closing class B fire door, it qualifies as a division wall for reduction of area.
3. Where communications are protected by a recognized water curtain, the value of P_i is 0.

* The factor for P_i depends on the protection for communicating party wall openings and the length and construction of communications between fire divisions. P_i shall be selected from this table. When more than one communication type exists in any one side wall, apply only the largest factor P_i for that side. When there is no communication on a side, $P_i = 0$. (Party wall means a division wall rated one hour or more with labeled class B fire doors on openings.)

† For more than 50 ft (15.3 m), $P_i = 0$.

‡ For unprotected passageways of this length, consider the two buildings as a single fire division.

Table 1-6 Needed fire flow for one- and two-family dwellings§

Distance Between Buildings		Needed Fire Flow	
ft	(m)	gpm	(L/sec)
More than 100	(more than 30.5)	500	(31.5)
31–100	(9.5–30.5)	750	(47.3)
11–30	(3.4–9.2)	1,000	(63.1)
Less than 11	(Less than 3.4)	1,400	(94.6)

§Dwellings not to exceed two stories in height.

The equation was based on the combustion of fuel being dependent on the available oxygen supply in the closed compartment and the vaporization of applied water into steam. The expansion ratio of water to steam was considered to assess the capability of vaporizing water to displace oxygen. Time-temperature curves for the fires were analyzed to determine optimal rates of water application for which steam generation was a maximum. The equation noted above has been revised by dividing by 200, 300, or 400 instead of 100 to relate the hazard imposed by the contents.

This equation is the easiest of the [four] to be discussed in this article. If the ceiling height in a room is approximately 10 feet, the equation reduces to:

$$G = \text{area of room (in square feet)} \div 10$$

A limitation of this equation is a result of the assumption that the entire space must be involved in fire. Thus, for a large, open warehouse or other noncompartmented building, use of this equation yields fire flows which will be quite large. (Milke 1980)

National Fire Academy Method[*]

The United States Fire Administration (USFA, www.usfa.dhs.gov) is a part of the Department of Homeland Security's Federal Emergency Management Agency. The USFA's mission is to reduce life and economic losses due to fire and related emergencies through leadership, advocacy, coordination, and support. The National Fire Academy (NFA, www.nfaonline.dhs.gov) promotes, for USFA, the professional development of the fire and emergency response community by delivering educational and training courses having a national focus.

One NFA training program, Command and Control Decision Making at Multiple Alarm Incidents (Q297), describes how to use a method for determining fire flow requirements. This method is a modification of the Iowa State University method. The NFA method uses a "quick calculation" formula as a tactical tool for use at the scene of an incident or for preplanning fire flow requirements for major structures. The basic quick calculation formula for a one-story building that is 100 percent involved is:

$$\text{Needed Fire Flow (NFF)} = \frac{\text{length} \times \text{width}}{3}$$

Length and width are in feet rounded to the nearest 10.
The calculated NFF is in gpm.

The basic formula is modified if the building is less than 100 percent involved as follows:

$$\text{Needed Fire Flow (NFF)} = \frac{\text{length} \times \text{width}}{3} \times \% \text{ involvement}$$

If more than one floor of a multistory building is involved, the fire flow calculated for each floor using the basic formula (modified by the percent involvement if less than 100 percent) is added to determine the total NFF for the building.

Interior and exterior exposures (exposure charge) also add to the needed fire flow. For interior exposure, add 25 percent of the 100 percent involvement figure determined by the basic formula for each floor above the fire floor for up to a maximum of five floors. For exterior exposure, add 25 percent of the 100 percent involvement figure determined by the basic formula for each side of the fire building that has an exposed building (within 30 feet) facing it.

---

[*] US Fire Administration, 16825 S. Seton Ave., Emmitsburg, MD 21727.

Illinois Institute of Technology Research Institute Method

The IITRI technique was developed from a survey. Data was collected from 134 fires in several occupancy types in the Chicago area to determine the water application rate needed for control as a function of fire area. Reported fires were of differing levels of magnitude, so not to concentrate solely on large-loss fires. Water application rates for the studied fires were calculated through a knowledge of length and diameter of hose used and calculated nozzle pressure.

Calculations.

The [needed] fire flow G, in gallons per minute, is calculated by one of the following equations:

$$\text{Residential Occupancies: } G = 9 \times 10^{-5} A^2 + 50 \times 10^{-2} A$$

$$\text{Nonresidential Occupancies: } G = -1.3 \times 10^{-5} A^2 + 42 \times 10^{-2} A$$

where A is the area of the fire in square feet.

These equations were obtained through a curve-fitting analysis of available data points on a graph. The investigation noted that tactical procedures can influence the application rate of water use greatly, e.g., interior versus exterior attack, leading off with large-diameter rather than small-diameter hose and similar concerns. As a supplement, 21 laboratory experiments on the use of manual streams to extinguish compartment fires also were reported for comparison purposes. Analysis of this experiment indicated that fire fighter training and comfort were key parameters in determining the amount and rate of water used, and the application rates of the 134 actual fires observed by IITRI were approximately double the rate used in the laboratory. (Milke 1980)

Comparison of Calculation Methods

Comparisons between the various techniques for computing fire flows are not easily made, because each situation to which the fire flow calculation is applied varies greatly. Comparisons are made here (Figures 1-3 and 1-4) to show the relative results obtained by the four different methods discussed for certain fairly typical situations.

The comparison here consists of two parts. First, a single incidence involving a building being evaluated for fire flow requirements is analyzed using the four methods. Next, a building of a fixed type of construction and configuration, but varying size, is analyzed using the four methods. This is done to illustrate how each method deals with the problem of relating the required fire flow to the size of the structure.

Part one—Three alternative construction scenarios. In this example, the subject building is 15,000 ft^2 (1,394 m^2) in size, one story, 12 ft high, and of ordinary construction. The building, occupied as a supermarket, is being analyzed to determine the fire flow required to control and contain fire within this structure. Because the ISO method involves evaluating the fire flow requirement for an adjacent structure, known as the *exposure building*, this situation is analyzed for three different exposure buildings. The exposure building variations used in these calculations are described in Figures 1-3 and 1-4. These two figures describe the situations being analyzed, provide a pictorial representation of the situations, and provide the fire flow calculations determined by the four methods discussed here.

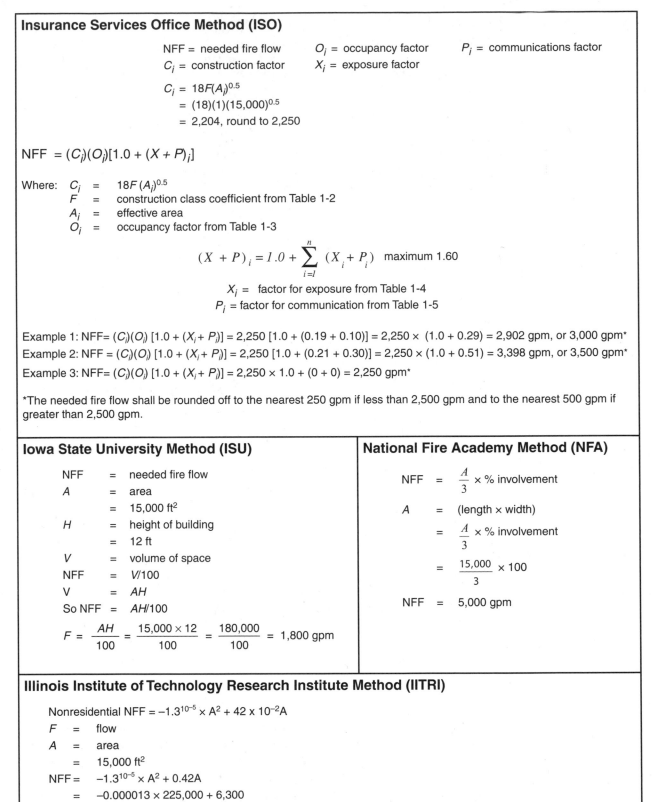

Insurance Services Office Method (ISO)

NFF = needed fire flow O_i = occupancy factor P_i = communications factor

C_i = construction factor X_i = exposure factor

$C_i = 18F(A_i)^{0.5}$
   $= (18)(1)(15,000)^{0.5}$
   $= 2,204$, round to 2,250

$NFF = (C_i)(O_i)[1.0 + (X + P)_i]$

Where: C_i = $18F(A_i)^{0.5}$
   F = construction class coefficient from Table 1-2
   A_i = effective area
   O_i = occupancy factor from Table 1-3

$$(X + P)_i = 1.0 + \sum_{i=1}^{n}(X_i + P_i) \quad \text{maximum 1.60}$$

X_i = factor for exposure from Table 1-4
P_i = factor for communication from Table 1-5

Example 1: NFF= $(C_i)(O_i)[1.0 + (X_i + P_i)]$ = 2,250 [1.0 + (0.19 + 0.10)] = 2,250 × (1.0 + 0.29) = 2,902 gpm, or 3,000 gpm*
Example 2: NFF = $(C_i)(O_i)[1.0 + (X_i + P_i)]$ = 2,250 [1.0 + (0.21 + 0.30)] = 2,250 × (1.0 + 0.51) = 3,398 gpm, or 3,500 gpm*
Example 3: NFF= $(C_i)(O_i)[1.0 + (X_i + P_i)]$ = 2,250 × 1.0 + (0 + 0) = 2,250 gpm*

*The needed fire flow shall be rounded off to the nearest 250 gpm if less than 2,500 gpm and to the nearest 500 gpm if greater than 2,500 gpm.

Iowa State University Method (ISU)

NFF   = needed fire flow
A   = area
   = 15,000 ft^2
H   = height of building
   = 12 ft
V   = volume of space
NFF   = V/100
V   = AH
So NFF   = AH/100

$$F = \frac{AH}{100} = \frac{15,000 \times 12}{100} = \frac{180,000}{100} = 1,800 \text{ gpm}$$

National Fire Academy Method (NFA)

NFF   = $\frac{A}{3}$ × % involvement

A   = (length × width)

   = $\frac{A}{3}$ × % involvement

   = $\frac{15,000}{3}$ × 100

NFF   = 5,000 gpm

Illinois Institute of Technology Research Institute Method (IITRI)

Nonresidential NFF = $-1.3^{10^{-5}} \times A^2 + 42 \times 10^{-2}A$
F   = flow
A   = area
   = 15,000 ft^2
NFF= $-1.3^{10^{-5}} \times A^2 + 0.42A$
   = $-0.000013 \times 225,000 + 6,300$
   = $-2,925 + 6,300$
NFF=   3,375 gpm

Figure 1-3   Comparison of fire flow calculations

NFF calculation example for four methods.

Nonresidential structure, ordinary construction
One story
Area = 15,000 ft² (1,394 m²)
Height = 12 ft
Adjacent structures (exposure building shown in Figure 1-4)

	X_i from Table 1-4	P_i from Table 1-5
Example 1	0.19	0.10
Example 2	0.21	0.30
Example 3	0.0	0.0

Part two—Various structure sizes. In the second part of the comparison, a one-story building of ordinary construction and nonresidential occupancy, but varying in size from 0 to 15,000 ft² (1,394 m²), is analyzed for each method to determine fire flow required for the various sizes of buildings.

Comparison. While there are no firm rules to follow when comparing the calculations derived from the four methods, there are some reasonable conclusions that can be made by comparing the two situations previously discussed.

The IITRI method generally yields the highest fire flow requirement. Generally, the ISO and ISU methods parallel one another, with the ISO method being somewhat, but not significantly, higher. This arises from a number of probable causes. First, the ISO method deals not only with the building presumed to be involved but it also considers the need to protect the exposure buildings. In addition, ISO factors the status of the fire department equipment and personnel experience and other variables into its calculations. The ISU method is a somewhat stylized approach. This method envisions that the water being supplied to fight a fire is applied in a theoretically ideal manner so as to obtain maximum effectiveness. Clearly this is not always an achievable situation. The NFA formula result depends on the percent involvement.

PRACTICAL LIMITS ON FIRE FLOW

Using an engine or hose company from a local fire department, which draws large amounts of water from the public water supply system, is not the preferred method of fire suppression. In many cases, an automatic fire suppression system, such as a sprinkler or a chemical system in combination with an alarm system, is more effective. In fact, a building developer who properly designs and installs a fire suppression system can do far more to protect life and property than a fire company can do with any amount of water delivered through the standard hose system. Fire sprinklers are intended to control a fire, not to completely extinguish it. Hose streams are almost always necessary. However, water from the public distribution system remains an important part of any fire suppression system.

Fire Flow Limits—Nonsprinklered Buildings

If the public water supply is to be used for fire suppression and a sprinkler system is not available, the supply available at a given point in the system is usually required to be no less than 500 gpm (32 L/sec) at a residual pressure of 20 psi (138 kPa). This represents the amount of water required for two standard hose streams on a given fire. Many professionals state that this is the minimum amount of water that can safely and effectively control any fire. Above that minimum, it is recommended that at any given point in the water distribution system, the system be able to provide the required

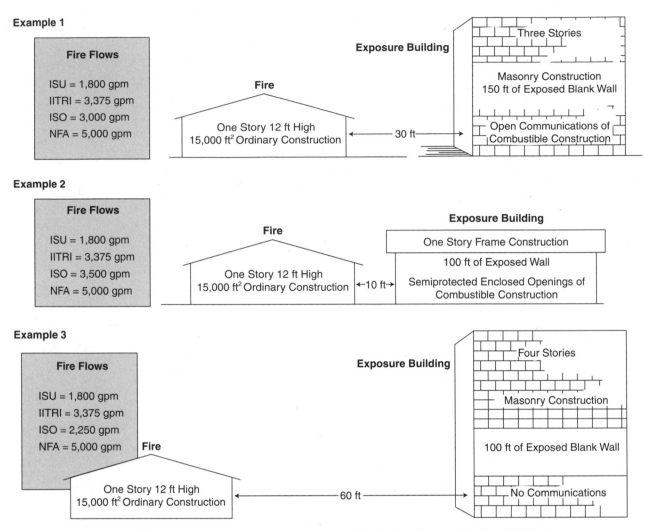

Example 1

Fire Flows

ISU = 1,800 gpm
IITRI = 3,375 gpm
ISO = 3,000 gpm
NFA = 5,000 gpm

Fire

One Story 12 ft High
15,000 ft² Ordinary Construction

←—30 ft—→

Exposure Building

Three Stories

Masonry Construction
150 ft of Exposed Blank Wall

Open Communications of
Combustible Construction

Example 2

Fire Flows

ISU = 1,800 gpm
IITRI = 3,375 gpm
ISO = 3,500 gpm
NFA = 5,000 gpm

Fire

One Story 12 ft High
15,000 ft² Ordinary Construction

←10 ft→

Exposure Building

One Story Frame Construction

100 ft of Exposed Wall

Semiprotected Enclosed Openings of
Combustible Construction

Example 3

Fire Flows

ISU = 1,800 gpm
IITRI = 3,375 gpm
ISO = 2,250 gpm
NFA = 5,000 gpm

Fire

One Story 12 ft High
15,000 ft² Ordinary Construction

←————60 ft————→

Exposure Building

Four Stories

Masonry Construction

100 ft of Exposed Blank Wall

No Communications

NOTE: Fire flows are from Iowa State University (ISU), Illinois Institute of Technology Research Institute (IITRI), Insurance Services Office (ISO), and National Fire Academy (NFA).

Figure 1-4   Comparison of fire flow calculations, including three exposure buildings

design flow as discussed earlier, or by using techniques adopted by responsible authorities. The ISO method is most likely to yield realistic requirements.

The NFF is a number used to evaluate the water system for fire insurance purposes. Actual flow used in fire fighting depends on the nature of the fire and how the fire department approaches the fire. There is no evidence that indicates a marginal shortfall in meeting the NFF can be related to an increase in loss. Inability of the distribution system to fully deliver NFF should not be considered a failure of the system.

In areas of a community with nonsprinklered buildings, the minimum fire flow provided to those areas is usually set at 500 gpm (32 L/sec) unless there are buildings that need higher fire flows. This is a community decision to be made by the community's governing body. If the water distribution system is serviced by a privately owned utility, some arrangement should be made by the governing body with that supplier to provide the required degree of protection.

Fire Flow Limits—Sprinklered Buildings

Installing fire sprinklers in a building can significantly reduce the NFF from the amount calculated by the methods presented earlier in this chapter. In such cases, the NFF is the sum of the sprinkler flow required at the base of riser plus a hose stream allowance. This is discussed further in chapter 5.

Exceptions to Fire Flow Limits

There are some exceptions to the required fire flow. For example, if a community has a large concentration of housing units with required fire flows not in excess of 1,500 gpm (95 L/sec) and a small number of properties require an increased level of flow (3,500 gpm [221 L/sec]), it would not make good economic sense to provide 3,500 gpm to those isolated properties. The community's governing body would be advised to simply develop ordinances and regulations that require those isolated properties to provide for their own private fire protection, to reduce the fire flow requirement by using a higher level of sprinkling, or to provide on-site storage and pumping capabilities to meet their own fire suppression needs.

There could be circumstances in which a community might arrange to deliver the upper limits of a required fire flow to an isolated building. For example, a single, large, high-hazard mercantile establishment, which provides most of the jobs in the community and produces most of the community's tax revenue, may receive the required fire flow from the community. By working with the building owner, adequate fire suppression could be provided. This might be achieved through sprinklers or some other means.

NONPOTABLE WATER SOURCES FOR FIRE FIGHTING

Numerous nonpotable water sources may be used as the primary or backup supply for fire protection. These sources may be divided into two major groups. One group comprises the nonpotable portion of a dual distribution system, which provides potable and nonpotable water to all or selected areas of a community.

Dual distribution systems for community water supplies are increasingly common. Diminishing supplies of high-quality resources and rapidly escalating costs of treating both potable water and wastewater are the two main causes.

As dual distribution systems become available, they will increasingly be used as water sources for fire suppression. When the systems are used for fire suppression, the design requirements are generally identical to those specified for potable systems in this manual, with the exception of special markings and fittings for public safety purposes. Detailed local regulations are still being developed. See AWWA Manual M24 *Dual Water Systems* for more information.

The other nonpotable water source consists largely of suction supplies, most frequently (but not exclusively) used as a source for private fire protection systems in accordance with NFPA Standard 1142, Water Supplies for Suburban and Rural Fire Fighting.

REFERENCES

City of Dallas Fire Flow Study. Spring 1983. *Sprinkler Quarterly*. Dallas, Texas.

Côté, A.E., and J.L. Linville, eds. 2003. *Fire Protection Handbook*. 19th ed. Quincy, Mass.: National Fire Protection Association.

Dual Water Systems. AWWA Manual M24. Denver, Colo.: American Water Works Association.

Fire Suppression Rating Schedule. 2003. Jersey City, N.J.: Insurance Services Office Inc.

Installation of Private Fire Service Mains. NFPA Standard 24-07. Quincy, Mass: National Fire Protection Association.

Installation of Sprinkler Systems. NFPA Standard 13-07. Quincy, Mass.: National Fire Protection Association.

Milke, J.A. 1980. How Much Water Is Enough? *The International Fire Chief* (March), pp. 21–24.

Sprinkler Systems in One- and Two-Family Dwellings and Manufactured Homes. NFPA Standard 13D-07. Quincy, Mass.: National Fire Protection Association.

Sprinkler Systems in Residential Occupancies Up To and Including Four Stories in Height. NFPA Standard 13R-07. Quincy, Mass.: National Fire Protection Association.

Water for Fire Fighting—Rate of Flow Formula. Bulletin 18. Ames, Iowa: Iowa State University Fire Extension Service.

Water Supplies for Suburban and Rural Fire Fighting. NFPA Standard 1142. Quincy, Mass.: National Fire Protection Association.

Chapter **2**

System Demand and Design–Flow Criteria

This chapter reviews the relationships between the various demands placed on a water distribution system and outlines recommended design- flow criteria. Before system demands or design–flow criteria can be discussed, a basic understanding of a water distribution system and its components is needed. Distribution systems are discussed along with the requirements placed on elements of the system.

METHODS OF DISTRIBUTION

Water is dispersed throughout the distribution system in several different ways, depending on local conditions or as other considerations may dictate. The common methods of distribution are discussed in the following sections.

Gravity Distribution

Gravity distribution is possible when the source of treated water is a lake or impounding reservoir at some elevation above the community. In this type of system, sufficient pressure is available to maintain water pressure in the mains for domestic and fire service. This is the most reliable method of distribution if the piping that leads from the source to the community is reliable and adequate in size. High pressure for fire fighting, however, may require the use of fire department pumpers or fire pumps.

Pumps With Elevated Storage

Through the use of pumps and elevated storage, excess water pumped during periods of low consumption is stored in elevated tanks or reservoirs. During periods of high consumption, the stored water supplements the water that is being pumped. This method allows fairly uniform rates of pumping. Consequently, this method is economical because smaller pumps can be used, and the pumps may be operated at their rated capacity. Because the stored water supplements the supply used for fires and system breakdowns, this method of operation is fairly reliable. However, fire pumpers must generally be used for higher fire pressures. It may be possible to close the valves leading to the elevated storage tanks and operate an auxiliary fire pump at the pumping plant during fire emergencies. A more complete discussion of storage is contained in chapter 3.

Pumps Without Storage

When pumps are used to distribute water and no storage is provided, the pumps force water directly into the mains. There is no other outlet for the water. Variable speed pumps or multiple pumps may be required to provide adequate service because of fluctuating demands. Another disadvantage is the fact that the peak power demand of the water plant is likely to occur during periods of high electric power consumption and thus increase power costs. An advantage of direct pumping is that a large fire-service pump may be used to increase residual pressure to any desired levels within the limits permitted by construction of the mains.

Systems with little or no storage should be provided with standby electrical generating capability or pumps directly driven by internal combustion engines. These standby generators and engines need to be tested routinely (e.g., several hours per week).

RATES OF WATER USE

Three historical or predicted water demand rates are involved in the discussion of fire protection. They are

- *Average daily demand*—the average of the total amount of water used each day during a one-year period.

- *Maximum daily demand*—the maximum total amount of water used during any 24-hour period. The ISO bases its calculations on the highest demand during the previous three years. (This number should consider and exclude any unusual and excessive uses of water that would affect the calculation.)

- *Maximum hourly demand*—the maximum amount of water used in any single hour of any day in a three-year period. It is normally expressed in gallons per day by multiplying the actual peak hour use by 24.

When specific data on past consumption are not available, a good rule of thumb is that maximum daily demand may vary from 1.5 to three times the average daily demand, while the peak hourly rate may vary from two to eight times the average daily rate. In small water systems, peaking factors may be significantly higher. Localized peaking factors may also be significantly higher where consumption patterns are highly seasonal in nature, where industrial applications have periodic extremes in flow rate demand, etc.

Design flow should be based on the maximum hourly demand or the maximum daily demand plus the fire flow requirement, whichever is greater. The distribution system should be designed to maintain a minimum pressure of 20 psi (138 kPa) at all points in the system under all conditions of design flow *(Recommended Standards for Water Works* 2003).

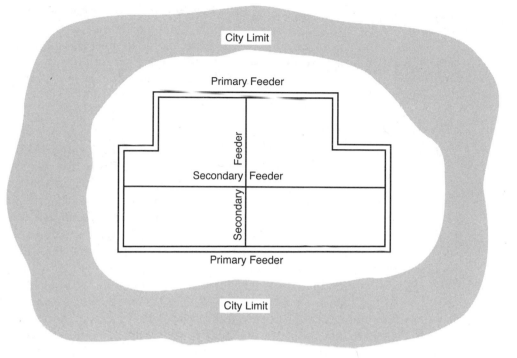

Figure 2-1   Typical small city distribution system

DISTRIBUTION SYSTEM APPURTENANCES

Piping and Valve Arrangement

A piping system serving the consumers in a small community is illustrated in Figure 2-1. The primary feeders, sometimes called the arterial mains, form the skeleton of the distribution system. They are located so that large quantities of water can be carried from the pumping plant to and from the storage tanks and distribution system.

Primary feeders should be arranged in several interlocking loops, with the mains not more than 3,000 ft (914 m) apart. Looping allows continuous service through the rest of the primary mains, even when one portion is shut down temporarily for repairs. Under normal conditions, looping also allows supply from two directions for large fire flows. Large feeders and long feeders should be equipped with blowoff valves at low points and air relief valves at high points. Valves should be placed so that a pipe break will affect water service only in the immediate area of the break.

The secondary feeders carry large quantities of water from the primary feeders to points in the system in order to provide for normal supply and fire fighting. They form smaller loops within the loops of the primary mains by running from one primary feeder to another. Secondary feeders should be spaced only a few blocks apart. This spacing allows concentration of large amounts of water for fire fighting without excessive head loss and resulting low pressure.

Smaller distribution mains form a grid over the area to be served. They supply water to fire hydrants and service pipes of residences and other buildings. Pipe sizing is usually based on the sum of the peak day water use plus fire flow requirements. In smaller mains, fire flows are usually much greater than normal or even peak day water use, so pipe sizing is almost completely controlled by fire flow requirements. In very large mains, because the additional flow from a fire is negligible when compared to peak day use, fire flows have little impact on pipe sizing.

Distribution piping should be sized and spaced to meet design flow. The minimum size of water mains for providing fire protection and serving fire hydrants is 6 in. (150 mm) in diameter. Typical values for distribution system piping are summarized in Table 2-1.

Hydrant Locations

All areas served by a distribution system should have hydrants installed in locations and with spacing convenient for fire department use. There are two methods for distributing hydrants in general use. The first method, known as the *area method*, is common in Canada. This method is fully described in *Water Supply for Public Fire Protection: A Guide to Recommended Practice*. In this approach, the area served by a hydrant is based on the NFF. For example, if the fire flow is 1,000 gpm (63.1 L/sec), each hydrant is expected to cover 160,000 ft^2 (14,864 m^2). As the NFF increases, the average area covered by a hydrant decreases.

The second method, the *linear method*, is outlined in section 614 of the *Fire Suppression Rating Schedule*. Summarized briefly, this procedure examines a representative location and, given the required fire flow at that location, determines if hydrants within 1,000 ft (305 m) of the location can provide the required fire flow at 20-psi (138-kPa) residual pressure. Note that the maximum gallon per minute flow is limited by the hydrant spacing. These limits are 1,000 gpm (63.1 L/sec) for hydrants within 300 ft (91.4 m), 670 gpm (42.3 L/sec) for hydrants from 301 to 600 ft (91.7 to 182.9 m), and 250 gpm (15.8 L/sec) for hydrants from 601 to 1,000 ft (183.2 to 304.8 m). Alternatively, a spacing criterion adopted by the authority having jurisdiction may be used. The water utility should discuss optimal fire hydrant spacing with the appropriate fire officials because the hose diameter and the hose capacity of fire trucks may limit spacing.

The pipe connecting the hydrant to the main is called the hydrant branch or lateral. Every lateral should contain a valve to enable the utility to isolate the hydrant for maintenance.

In addition, fire department use normally requires a maximum lineal distance between hydrants along streets in congested areas of 300 ft and of 600 ft (91.4 and 182.9 m) for light residential districts. Good practice calls for hydrants at intersections; in the middle of long blocks, particularly when required flows exceed 1,300 gpm

Table 2-1   Values commonly used in distribution piping

	Appurtenance	Typical Minimum Values	
Lines	Smallest pipes in network	6 in.	(150 mm)
	Smallest branching pipes (dead ends)	8 in.	(200 mm)
	Largest spacing of 6-in. grid (8-in. pipe used beyond this value)	600 ft	(183 m)
	Smallest pipes in high-value district	8 in.	(200 mm)
	Smallest pipes on principal streets in central district	12 in.	(300 mm)
	Largest spacing of supply mains or feeders	3,000 ft	(914 m)
Valves	Spacing in single- and dual-main systems:		
	Largest spacing on long branches	800 ft	(245 m)
	Largest spacing in high-value district	500 ft	(152 m)
Distribution system pressure		20 psi	(138 kPa)

(82 L/sec); and near the end of long dead-end streets. Hydrants should be required within large properties accessible to fire apparatus. Planning hydrant locations should be a cooperative effort between the water utility and the fire department.

The actual distance between hydrants is determined somewhat on the amount of hose that the local fire department carries; the numbers vary from one fire department to another. In instances where light residential areas are protected by residential sprinkler systems, there may be merit in assessing reduction in line sizes or in hydrant spacing.

SYSTEM EVALUATION AND DESIGN

The adequacy of the existing water distribution system is usually evaluated using fire hydrant flow tests. AWWA Manual M17 *Installation, Field Testing, and Maintenance of Fire Hydrants* describes how to conduct such tests. Analysis of proposed systems and improvement for correcting deficiencies in existing systems are usually evaluated with water distribution system models.

Except for systems solely dedicated to fire protection, the fire protection capability of a water distribution system cannot be evaluated independently of the system's ability to meet other demands. This type of analysis is usually conducted as part of the planning for land development or in the typical 5- or 10-year master plan prepared by most utilities. The distribution system's ability to operate adequately under fire demands is an important feature of such planning studies. Improvements should be designed not only to meet current needs but also to fit into an integrated long-range plan for the utility. More information is available from agencies involved in fire protection listed in appendix B. Adequacy of water supply is one of the factors used to develop insurance rates for properties. Appendix C focuses on water supply and fire insurance ratings.

DETERMINING DESIGN FLOW

The first step in evaluating or designing a fire protection system is to determine design flow for the area of concern and/or for several representative areas, if an overall evaluation of a system is to be adopted. Design flow is usually calculated as follows:

1. Determine the average daily demand.

2. Determine the maximum daily demand or estimate it from the average daily demand.

3. Determine the maximum hourly demand or estimate it from the average daily demand.

4. Determine the required fire flow using one of the methods discussed in chapter 1 for nonsprinklered properties or another method approved by the authority with jurisdiction.

5. If the properties in question are to be sprinklered, determine the required flow from the sprinkler requirement plus fire stream allowance.

6. Select an appropriate design flow for the system at hand. Normally, this is either the sum of the required fire flow for the most stringent situation (nonsprinklered or sprinklered properties in the area in question) plus the maximum daily demand or the maximum hourly demand, whichever is greater.

Good judgment must be exercised at this point. For example, if the property in question is a manufacturing facility that uses large amounts of water for only eight hours during one shift per day, five days a week, with relatively little domestic or other use, the maximum hourly demand would be far in excess of what one would normally anticipate, given the average or maximum daily demand. In this case, the system serving the facility should be designed to provide the required flow plus the maximum hourly flow. It would be foolish and irresponsible to consider treating a fire at this facility on the basis of an average daily demand that has no real meaning in this particular context.

7. Evaluate or design the system using the design flow as previously determined to ensure that the reliability considerations discussed in chapter 4 are adequately dealt with.

FLOW METERING

The decision of whether or not to meter dedicated fire lines is at the discretion of the local water utility and is based on the tradeoff between the extra cost of reading the meters and additional head loss versus improved accountability for water used.

Meters installed for a dedicated or dual purpose fire line should be designed for that purpose. Note that AWWA C703 currently defines design requirements only for meters in fire service lines 3 in. and larger, while extension of this standard to cover meters 2 in. and smaller is currently under review. In dedicated fire lines, alarms, detector check valves, and flowmeters should be approved by the local fire official.

Downsizing fire line meters on dedicated or dual purpose lines can increase the accuracy of metering but should not interfere with supplying adequate fire flows. If a wide range of flow is expected, it is usually preferable to install an approved fire line meter assembly, assembled and tested, and complete with a bypass meter to register low flows.

STANDBY CHARGES FOR FIRE PROTECTION SYSTEMS

Water utilities can levy one-time capital recovery fees or annual standby charges for fire protection systems. These charges should be based on the actual cost to provide the service.

WATER DISTRIBUTION ANALYSIS TECHNIQUES

Water system sizing today is generally done using hydraulic models of the distribution system to simulate a wide range of critical flow events, including fire flows. These models enable the engineer to examine a broad range of events that may determine pipe, pump, or tank sizing.

Any model used for distribution system sizing must be calibrated to match the performance of the existing system not only during average flow conditions but also during high flow events. Calibrating the model to agree with the results of hydrant flow tests ensures that the model is performing correctly. Procedures for conducting hydrant flow tests are described in AWWA Manual M17 *Installation, Field Testing, and Maintenance of Fire Hydrants*.

Modern pipe network models make it possible to simulate fire flows at any point in the water distribution system and to determine the available fire flow for any set of conditions (e.g., tank water levels and pump operations). Applying models for distribution system design is described in AWWA Manual M32 *Computer Modeling of Water Distribution Systems*.

REFERENCES

Ballou, A.F. 1938. Hydrant Connections and the Control of Dual Mains. *NEWWA Journal* 52:81.

Cesario, A.L. 1995. *Modeling, Analysis, and Design of Water Distribution Systems*. Denver, Colo.: American Water Works Association.

Cold-Water Meters—Fire Service Type. AWWA Standard C703. Denver, Colo.: American Water Works Association.

Computer Modeling of Water Distribution Systems. AWWA Manual M32. Denver, Colo.: American Water Works Association.

Fire Suppression Rating Schedule. 2003. Jersey City, N.J.: Insurance Services Office Inc.

Installation, Field Testing, and Maintenance of Fire Hydrants. AWWA Manual M17. Denver, Colo.: American Water Works Association.

Recommended Standards for Water Works. 2003. Great Lakes–Upper Mississippi River Board of State and Provincial Public Health and Environmental Managers. Health Education Services. Albany, N.Y.: State Health Department.

Water Supply for Public Fire Protection: A Guide to Recommended Practice. 1981. Toronto, Ont.: Fire Underwriters Survey.

Chapter **3**

Distribution System Storage

An extremely important element in a water distribution system is water storage. System storage facilities have a far-reaching effect on a system's ability to provide adequate fire protection. The two common storage methods—ground storage and elevated storage—are discussed in this chapter. Emphasis is placed on the relative merits of both methods.

FUNCTIONS OF DISTRIBUTION STORAGE

Water storage is designed into water distribution systems for a variety of reasons. Storage can be designed to equalize pressures and minimize pressure transients across a distribution zone, provide emergency supply for fire fighting operations, and better distribute peak flows while optimizing pumping power consumption and costs. Storage can be designed to distribute and store water at one or more locations in the service area that are closer to the user.

Advantages

The principal advantages of distribution storage are that storage equalizes demands on supply sources, production works, and transmission mains. As a result, the sizes or capacities of these elements can be smaller. Additionally, system flows and pressures are improved and stabilized to better serve the customers throughout the service area. Finally, reserve supplies are provided in the distribution system for emergencies, such as fire fighting and power outages.

Meeting System Demands and Required Fire Flow

The location, capacity, and elevation (if elevated) of distribution storage are closely associated with system demands and the variations in demands that occur throughout the day in different parts of the system. System demands can be determined only after a careful analysis of an entire distribution system. However, some general rules serve as a guide to such analysis. Table 3-1 lists daily and hourly variations for a typical city and the resultant storage depletion. Such data are of great assistance in determining required storage capacities. Of course, each system has its own requirements.

Rarely can distribution storage be economically justified in an amount greater than what will take care of normal daily variation and provide the needed reserve for fire protection and minor emergencies. In systems of moderate size, the amount of water storage available for equalizing water production is 30–40 percent of the total storage available for water pressure equalization purposes and emergency water supply reserves. In normal system operations, some water from storage should be used each day not only to maintain uniformity in production and pumping, but also to ensure circulation of the stored water and prevent the loss of disinfectant residual. Storage in the distribution system is normally brought to full capacity each night and is increased during low demand periods of the day.

It is normally more advantageous to provide several smaller storage units in different parts of the system than to provide an equivalent capacity at a central location. Smaller pipelines are required to serve decentralized storage, and other things being equal, a lower flow line elevation and pumping head result.

Table 3-1   Water use and storage depletion for maximum day in a typical city

Hr	Ratio of Hourly Demand Rate to Maximum Day Demand Rate*	Hourly Variation in Distribution Storage Reserve *mil gal*	Cumulative Storage Depletion *mil gal*
7–8 a.m.	1.00	−0.00	0.00
8–9	1.10	−0.10	0.10
9–10	1.25	−0.25	0.35
10–11	1.28	−0.28	0.63
11–12	1.20	−0.20	0.83
12–1 p.m.	1.18	−0.18	1.01
1–2	1.16	−0.16	1.17
2–3	1.10	−0.10	1.27
3–4	1.00	−0.00	1.27
4–5	1.08	−0.08	1.35
5–6	1.15	−0.15	1.50
6–7	1.30	−0.30	1.80
7–8	1.60	−0.60	2.40
8–9	1.40	−0.40	2.80
9–10	1.25	−0.25	3.05[†]
10–11	0.90	+0.10	2.95
11–12	0.85	+0.15	2.80
12–1 a.m.	0.70	+0.30	2.50
1–2	0.60	+0.40	2.10
2–3	0.50	+0.50	1.60
3–4	0.50	+0.50	1.10
4–5	0.50	+0.50	0.60
5–6	0.60	+0.40	0.40
6–7	0.80	+0.20	0.20

* Average day, 16 mil gal; maximum day, 25 mil gal; constant hourly supply rate (at maximum day demand rate), 24 mgd or 1 mil gal/hr.

† Maximum storage depletion.

Minimum storage requirements are sometimes specified by regulatory agencies in terms of gallons of storage per capita or per connection. These numbers are usually more conservative than the volume that would be required by adding equalization storage plus fire storage, especially for larger pressure zones. There may be a reevaluation of these storage requirements as public health officials become more concerned about the deterioration of water quality in storage tanks.

In cases where water utilities (or pressure zones) are interconnected, it may be possible for one utility to provide some or all of the storage required for the second utility if (1) there is adequate hydraulic carrying capacity in the pipes between the two utilities and the second utility (or pressure zone) operates at the same or lower hydraulic grade line, or (2) reliable pumping is in place.

ELEVATED AND GROUND STORAGE

Storage within the distribution system is normally provided in one of two ways: elevated storage or ground storage with high-service pumping. Elevated storage provides the best, most reliable, and most useful form of storage, particularly for fire suppression.

In this section, *ground tanks* refers to unpressurized tanks that have a water level below the hydraulic grade line so that water must be pumped from the tank. Some individuals refer to this type of tank as pumped storage. Ground level tanks located at high elevations are considered elevated tanks in this section.

Ground Storage

Because water kept in ground storage is not under pressure, it must be delivered to the point of use by pumping equipment. This arrangement limits system effectiveness for fire suppression in three ways: (1) there must be sufficient excess pumping capacity to deliver the peak demand for normal uses as well as any fire demands, which requires an investment in pumping capacity; (2) standby power sources and standby pumping systems must be maintained at all times because the system cannot function without the pumps; and (3) the distribution lines to all points in the system must be larger to handle peak delivery use plus fire flow, no matter where a fire might occur.

When ground storage is used in areas of high fire risk, the energy needed to deliver the water is lost on the initial delivery of water to the tank. The water must be repumped and repressurized with the consequent addition of more standby generators and more standby pumps. In addition, the system's high-service pumps must be either variable speed or controlled by discharge valves to maintain constant system pressures. This equipment is expensive, uses additional electric power, and requires extensive operation and maintenance. Frequently, the long-term energy costs and the additional capital costs for pumps, generators, and backup systems significantly increase the cost of a ground storage system.

Elevated Storage

Properly sized elevated water tanks provide dedicated fire storage and are used to maintain constant system pressure.

Domestic water supplies are regularly fed to the system from the top 10 or 15 ft (3 or 4.6 m) of water in the elevated tanks. As the water level in the tank drops, the tank controls call for additional high-service pumps to start in order to satisfy the system demand and to refill the tanks. The remaining water in the tanks (30 to 75 percent) is normally held in reserve as dedicated fire storage. This reserve automatically feeds into the system as the fire flow demand and the domestic use at a specific time exceed the capacity of the system's high-service pumps.

PUMPING FOR DISTRIBUTION STORAGE

The two types of distribution storage—ground and elevated—have two types of pumping systems. One is a direct pumping system, in which the instantaneous system demand is met by pumping with no elevated storage provided. The second type is an indirect system in which the pumping station lifts water to a reservoir or elevated storage tank, which floats on the system and provides system pressure by gravity.

Direct Pumping

Direct pumping usually involves variable speed pumps or multiple pumps with control based on system pressure. Hydropneumatic tanks at the pumping station provide some storage. These tanks permit the pumping station pumps to start and stop based on a variable system pressure preset by controls operating off the tank.

If direct pumping from ground tanks is used such that the water level in the ground tank is significantly lower than the hydraulic grade line elevation near the tank, considerable energy is wasted by filling the tank and then repumping the water.

Indirect Pumping

In an indirect system the pumping station is not associated with the demands of the major load center. It is operated from the water level difference in the reservoir or elevated storage tank, enabling the prescribed range of water levels in the tank to be maintained. Most systems have an elevated storage tank or a reservoir on high ground floating on the system. This arrangement permits the pumping station to operate at a uniform rate, with the storage either making up or absorbing the difference between station discharge and system demand.

LOCATION OF STORAGE

It is generally best to locate storage on the opposite side of the major load center from the main source of water (e.g., pumping station or well). In this way, water can flow toward the most likely fire location from more than one direction, enabling the utility to use somewhat smaller pipes. Tank location is often dictated by terrain and land availability. As a rule, elevated storage placed near the main pumping station is the least helpful for either fire fighting or equalization.

Figures 3-1 through 3-3 show three scenarios for distribution system storage. System A with no storage is used in situations where storage is impractical and the distance from the pumping station to the major load center is small. System C generally results in lower pumping heads and smaller pipe sizes than System B. The decision on storage should be based on a thorough hydraulic analysis.

STORAGE RESERVOIR OVERFLOW LEVEL

The selection of overflow elevation for distribution storage is important because it determines the layout of the transmission and distribution system. In most cases, the hydraulic grade line has already been determined by the overflow elevations on existing tanks. However, when given the opportunity to lay out a new system or pressure zone, the distribution system engineer must look at the long-term impacts of tank overflow elevation.

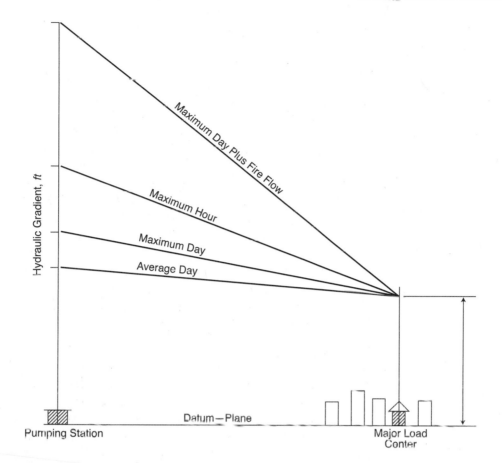

Figure 3-1 System A—hydraulic gradient with no storage

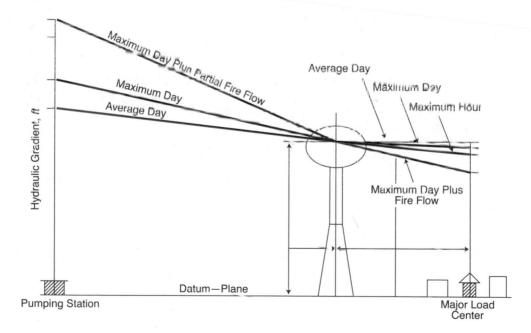

Figure 3-2 System B—hydraulic gradients with storage between pump station and load center

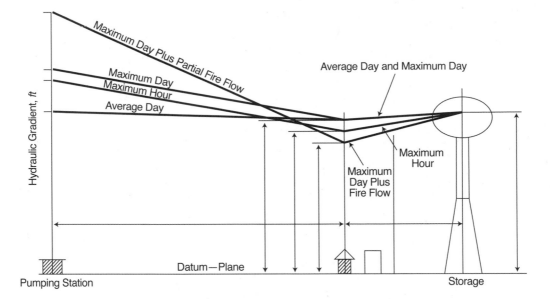

Figure 3-3   System C—hydraulic gradients with storage beyond load center

Overflow elevations in adjacent pressure zones should differ by 60 to 120 ft (18 to 36 m). Too large a difference between hydraulic grade lines in adjacent pressure zones results in problems providing reasonable pressures near pressure zone boundaries. Too small a difference results in excessive numbers of pumps and pressure-regulating valves.

The ground elevation that can be served by gravity from a storage reservoir should be examined before selection of overflow elevation. This can be done by determining the highest and lowest elevation that can be served given the overflow elevation and bottom elevation of the operating range.

REFERENCES

Côté, A.E., and J.L. Linville, eds. 2003. *Fire Protection Handbook*. 19th ed. Quincy, Mass.: National Fire Protection Association.

Fair, G.M., et al. 1966. *Water and Waste-water Engineering*. New York: John Wiley and Sons Inc.

Steel, E.W., and T.G. McGhee. 1991. *Water Supply and Sewerage*. New York: McGraw-Hill Book Company.

Water Distribution Operator Training Handbook. 2005. Denver, Colo.: American Water Works Association.

Chapter **4**

Adequacy and Reliability of Distribution Systems

A water supply system is considered to be fully adequate if it can deliver the required fire flows to all points in the distribution system with the consumption at the maximum daily rate (average rate on maximum day of a normal year). When delivery is also possible with the most critical limiting component out of service for a specified length of time, depending on the type of component, the system is considered to be reliable.

DETERMINING LEVEL OF RELIABILITY

It is virtually impossible to quantify the reliability of a particular point in a distribution system, much less the reliability of the entire system. Consequently, this manual can only recommend a qualitative level of reliability. Good engineering judgment must be used when specifying system components and the number of components needed to meet reliability standards.

The rationale for selecting a level of reliability and, in turn, the rationale for choosing system components and their number and location within a system are primarily socioeconomic. A water utility cannot be expected to perform a detailed cost–benefit analysis of potential fire loss. Allocating costs are impossible. Therefore, the sociological impact of fire losses should be considered, as well as life safety issues. The willingness of customers to pay for water system improvements for fire protection is also a consideration, especially when the new customer must pay for some or all of the improvements necessary for fire protection.

There is no such thing as a water system that is 100 percent reliable. There is always the possibility, for example, that the main in front of a building will rupture hours before the building catches fire. Water utilities should not guarantee that pressure or flow will be provided.

Providing Required Fire Flow

Distribution system components are often taken out of service for maintenance. Sometimes system components fail. In response to these occurrences, utilities should construct their distribution systems with loops, backup pumps, backup power supplies, and storage tanks. This is done so that if any component fails or is out of service the effect on available fire flows will be negligible.

Components out of service. Water system components are generally out of service for short periods of time, so the probability of a component being out of service when a fire occurs is low. Similarly, when a component is out of service for maintenance purposes, the utility has selected the time for this maintenance. For example, it is unlikely that a water tank will be painted while a large main leading to that part of the distribution system is being cleaned. The exceptions to this are rare but notable.

Fortunately, fires that severely stress a distribution system occur only a few times a year in large systems and only once every few years in small systems. Therefore, the probability of a major fire occurring while more than one water system component is out of service is so low that the utility should not be expected to meet required fire flow at such times. However, the system should be designed to provide some water at 20 psi (138 kPa). The fire department should be trained not to allow contamination because of low pressures in mains and service lines, even if some component is out of service during a fire. Automatic shutoff equipment is available on some fire trucks to prevent this problem.

Water supply facilities can be divided into major transmission facilities and local distribution lines. Because major transmission facilities serve large areas, the probability of a fire occurring in an area served by the system component while that component is out of service is much greater than the probability of a fire occurring in an area served by a local distribution line while that line is out of service. For example, if there are 20 houses on the 1700 block of Elm Street, it is unlikely there will be a fire on Elm Street during any day when that main is out of service. On the other hand, if one half of the water to a town is provided through the River Street main, it is quite likely that a fire will occur in an area affected by a shutdown in the River Street main.

The distinction between major transmission facilities and local distribution lines is not sharp. In small utilities, an 8-in. (200-mm) pipe may be part of the major transmission system, while in a large utility, a 12-in. (300-mm) pipe may be considered a local distribution line. The distinction is not based on size but rather on function. A main serving a large number of customers or an important customer is a major transmission main regardless of its size.

Looped systems. Loops are provided (1) to enable water to reach a customer (or fire) from more than one direction and hence greatly reduce head loss, and (2) to provide an alternative path for water to reach a customer in case one pipe is out of service. If a local distribution line is looped, it should not be expected to provide required fire flow if a portion of the loop is out of service. (The actual size of the lines in the loops must be determined by hydraulic analysis.) The rationale for this is that it is highly unlikely that one of the pipes in the loop will be out of service when a fire occurs and the lines are looped. Even if a portion of a loop is out of service, some water could be delivered to the fire. Local distribution lines need only be sized to provide required fire flow when all components are operating. Some municipalities may require that pipes be sized so that the entire fire demand must pass through one side of a loop. However, especially for neighborhood distribution pipes, the ability of the distribution system to convey fire flows is usually evaluated with all pipes in service.

Major System Components

Specifying the reliability required for major transmission mains, pump stations, and storage tanks is difficult because the outage of any one of these components can affect a large area. Furthermore, the chance that a fire will occur during an outage of one of these components is considerably higher than during an outage of local lines. The rule for determining reliability of these components is that all required fire flows be delivered downstream, even when one of these major components is out of service. In the case of pumps, the impact of loss of power to the pumps for an extended period of time needs to be considered and planned for.

For large systems, because water storage capacity to meet required fire flows may represent a small portion of the distribution system capacity, it does not represent a large extra cost to utilities. However, providing complete reliability in local distribution lines (e.g., size local distribution mains to provide required fire flow even if part of the loop is out of service) would be extremely costly.

APPLICATION OF RELIABILITY CONSIDERATIONS

Reliability considerations are important for distribution system design as performed by the utility, and for individual fire protection evaluation as performed by a building (or subdivision) designer or insurance professional. Each involves a different perspective and is addressed separately in the following sections.

Distribution System Design and Operation

In designing or evaluating a distribution system, the utility engineer should first separate components into local and major components. Local components should have some standby capability, but the engineer does not need to perform detailed analysis of the effects of each component being out of service. For the major components, the engineer should explicitly consider what will occur if the component is out of service and calculate the fire flows that can be delivered for each major outage. If fire flows cannot be met, the utility should consider upgrading the system. The upgrade may include some combination of increasing pipe size during design, installing additional looping pipes, installing additional tanks and pumps, or rehabilitating old pipes that have lost their carrying capacity.

The manner in which the utility operates and maintains the system is also very important in providing required fire flows. Because the system is most vulnerable to problems when components are out of service, the utility needs to carefully plan for these outages. This may include testing backup pumps before they are needed or inspecting and resetting pressure-reducing valves. A utility with a computer model of its system can simulate major fire flows with a component out of service to determine the impacts and to ensure that steps are taken to minimize the effects before the outage. Valves needed to isolate the out-of-service component should be located and routinely operated before they are needed.

A utility can minimize the effects of emergency shutdowns by confining the outage to the smallest possible area. To do this, the utility must have numerous valves, maintenance personnel who know the location of these valves and who have ready access to records of those locations, valve boxes that are free from debris, and valves in good operating condition. Only frequent exercising of valves can help ensure this.

If service is to be interrupted during a pipe cleaning operation or for other long-term interruptions, there are several ways to provide some reasonable fire protection. These include installing a temporary aboveground hose or other bypass lines, and developing a plan to quickly bring the line being cleaned or the device out of service back into service for fire protection, even though it has not been disinfected. If a pipe breaks near a fire, it may be advisable to delay isolating and repairing the break until after the fire. This is a judgment decision dependent on the relative damages of the fire compared to those of the break.

If there is concern over the effect that a pipe break may have on flow delivered to a fire, it is possible to simulate the performance of the distribution system during the outage by using a pipe network model. A model can be used to simulate pipes being taken out of service by closing valves. This is done by deleting or closing the model pipe links corresponding to pipes that have been shut off. Depending on the placement of valves used to isolate a pipe break, a pipe break does not simply take a single pipe link out of a model but instead takes out a distribution system segment, which may include portions of adjacent pipes and nodes.

Individual Fire Protection Evaluation

Individual fire protection systems and fire sprinklers can make it considerably easier to fight fires. As previously mentioned, utilities should be careful not to guarantee that any pressure will be available for the individual protection system. It is up to the owner (or the qualified engineer working for the owner) to determine a reasonable worst-case pressure in the utility's system to use as a basis for sizing individual fire protection systems.

Conducting a fire hydrant flow test near the location where the individual system will connect with the utility's system is a good way to assess the amount of water that can be delivered to a site under a given set of conditions for the existing water system.

RELIABILITY OF MAJOR SYSTEM COMPONENTS

The reliability of a water supply system depends on the reliability of all the system components within that system. Major system components discussed here include storage, pumping, buildings and plant, system operations, and emergency services.

Storage

Storage facilities usually reduce the requirements of any system components through which supply has already passed. Because storage levels usually fluctuate, the normal daily minimum that is maintained should be considered as the amount available for fire suppression. Because system pressure decreases when water is drawn from storage that floats on the system, only the portion of normal daily minimum storage that can be delivered at a residual pressure of 20 psi (138 kPa) at the point of use is considered to be available. In addition to the quantity available, the rate of water delivery to the system from storage during the fire flow period is critical.

Storage reservoirs provide reliability only when they are in service. Removing a tank from service for maintenance, such as painting or coating, reduces system reliability. Consideration should be given to selecting tanks that need not be drained or that are provided with cathodic protection to increase the period between maintenance.

Pumping

Reliability of pumping capacity. Pumping capacity should be sufficient, in conjunction with storage when the single most important pump is out of service, to maintain the maximum daily demand rate plus the maximum required fire flow at required pressure and duration. (The most important pump is typically the one of largest capacity, depending on how vital it is to maintaining flow to the distribution system.) For systems without an excessive number of pumps, the probability of a serious fire occurring when two pumps are out of service is low.

For pump stations that serve a small area with fire demands much greater than domestic demands (i.e., fire pump much larger than domestic pump), many utilities choose not to provide a backup fire pump because of the low probability of a fire with the fire pump out of service. For the same reason, the standby generator may be sized to meet peak domestic demands only.

For the system to be considered adequate, the remaining pumps, in conjunction with storage facilities, should be able to provide required fire flows for the specified durations at any time during a period of five days with consumption at the maximum daily demand. The normal minimum capacity of elevated storage located on the distribution system and storage with low-lift pumps should be considered. The rate of flow from such storage must be considered in terms of any limitation of water main capacity. The availability of spare pumps that can quickly be installed may be taken into consideration along with pumps of compatible characteristics that may be valved from another service to assist in fighting a fire.

Power supply. When feasible, electric power should be provided to all pumping stations and treatment facilities by two separate lines from different directions. The power supplies to pumps should be arranged so that a failure in any power line or the repair or replacement of a transformer, switch, control unit, or other device will not prevent the delivery of required fire flows. The extent to which standby power is required is described in various fire protection codes and state or provincial water supply regulations.

Underground power lines and lines with no other customers en route provide the highest level of reliability. Use of transmission lines by other consumers increases the possibility of a power interruption or deterioration of power characteristics.

Overhead power lines, which are more susceptible to damage and interruption than underground lines, introduce a degree of unreliability because of their location and construction. Consideration should be given to weather conditions, including lightning, wind, sleet, and snow storms in the area; the type of poles or towers and wires; the nature of the terrain traversed; the effect of earthquakes, storm events, forest fires, and floods; the provision of lightning and surge protection; the extent to which the system is dependent on overhead lines; and the ease of repairs.

The possibility of power system or network failures affecting large areas should also be considered. In-plant auxiliary power or internal combustion driver standby pumping are appropriate solutions to power failures. This is especially true in small plants where high pumping capacity is required for fire protection.

Fuel supply. Usually tanks capable of holding at least a five-day supply of fuel for internal combustion engines or boilers used for regular domestic supply should be provided. If long delivery times, poor road conditions, climatic conditions, or other circumstances could interrupt delivery for longer than five days, a larger storage facility for the fuel should be provided. Any gas supply should be from two independent sources or from duplicate gas-producing plants with gas storage sufficient for 24 hr. Unreliability of a regular fuel supply may be offset in whole or in part by the use of an alternate fuel or power supply.

Buildings and the Water Treatment Plant

Buildings and structures. Pumping stations, treatment plants, control centers, and other important structures should be located, constructed, arranged, and protected so that damage by fire, flooding, or other causes is minimal. These structures should contain no combustible material in their construction. If hazards are created by equipment or materials located within the structure, the hazardous section should be suitably separated by fire-resistant partitions or fire walls.

Preferably no structure should have fire exposure. If exposures exist, suitable protection should be provided. Electrical wiring and equipment should be installed in accordance with the National Electrical Code or the Canadian Electrical Code. All internal hazards should be properly safeguarded in accordance with good practice. Private in-plant fire protection should be provided as needed.

Miscellaneous system components, piping, and equipment. Steam piping, boiler feed lines, fuel piping (gas or oil lines to boilers as well as gas, oil, or gasoline lines to internal combustion engines), and air lines to wells or control systems should be arranged so that a failure in any line or the repair or replacement of valves, fuel pumps, boiler-feed pumps, injectors, or other necessary devices will not prevent the delivery of the required fire flows. This delivery, in conjunction with storage, should be available for the specified duration at any time during a period of five days with consumption at the maximum daily demand.

Plants should be arranged for effective operation. Consider the following factors: difficulty of making repairs, danger of flooding as a result of broken piping, susceptibility to damage by spray, reliability of priming and chlorination equipment, lack of semiannual inspection of boilers or other pressure vessels, dependence on common nonsectionalized electric bus bars, poor arrangement of piping, and poor condition or lack of regular inspections of important valves. Also consider factors affecting the operation of valves or other devices necessary for fire service, including the design, operation, and maintenance of pressure-regulating valves, altitude valves, air valves, and other special valves or control devices; provision of power drives; location of controls; and susceptibility to damage.

Reliability of the treatment plant is likely to be influenced by removing at least one filter or other treatment process unit from service; reducing filter capacity by turbidity, freezing, or other conditions; the need for cleaning basins; and the dependability of power for operating valves, backwash pumps, mixers, and other equipment.

OPERATIONS

Reliability of supply system operations and adequate response to emergency or fire demands are essential. Instrumentation, controls, and automatic features should be arranged with this in mind. Failure of an automatic system to maintain normal conditions or to meet unusual demands should sound an alarm so that remedial action can be taken. System operators should be competent, adequate, and continuously available, as required to maintain domestic and fire service.

Hydrant Testing and Maintenance

Hydrant testing and maintenance should be performed on a routine basis either by the water utility or the fire department. Maintenance and testing procedures are described in AWWA Manual M17 *Installation, Field Testing, and Maintenance of Fire Hydrants*. Records should be maintained and analyzed periodically to determine if there is any long-term loss in distribution system carrying capacity.

While the exact responsibilities of the water utility and the fire department may vary widely from one location to another, it is important for the two groups to clearly understand each other's role in hydrant maintenance and other issues. Good communication between water officials and fire officials is important not only for hydrant maintenance but in all areas of mutual concern. For example, the water utility should communicate to the fire department information about which hydrants can provide high flows and which ones are marginal by color coding hydrants or by providing maps of the distribution system.

Emergency Services

Emergency crews, provided with suitable system records, transportation, tools, and equipment, should be continuously on duty in the larger systems and readily available on call in small systems. Fire alarms should be received by the utility at a location where someone who can take appropriate action is always on duty. It may be necessary to place additional equipment in operation, operate emergency or special valves, or adjust pressures. Receipt of alarms may be by fire alarm circuit, radio, outside alerting device, or telephone. Where special operations are required, the alarm service should be equivalent to that needed for a fire station. Emergency utility crews should assist the fire department in making the most efficient use of the water system and ensure the best possible service in the event of a water main break or other emergency.

REFERENCES

Specific Commercial Property Evaluation Schedule. 1990. Jersey City, N.J.: Insurance Services Office Inc.

Installation, Field Testing, and Maintenance of Fire Hydrants. AWWA Manual M17. Denver, Colo.: American Water Works Association.

Chapter **5**

Automatic Fire Sprinkler Systems

Automatic sprinklers have been the most important single system for automatic control of hostile fires in buildings for more than a century . . . Among the benefits of automatic sprinklers is the fact that they operate directly over a fire. Smoke, toxic gases, and reduced visibility do not affect their operation. In addition, much less water is used because only those sprinklers fused by the heat of the fire operate, especially if the building is compartmented. (Côté 2003)

This statement points out that of all the tools available to facilitate and promote fire protection, none offers such a wide variety of benefits to the building owner, developer, fire service, water purveyor, and the general public as the widespread use of automatic sprinkler systems.

In this chapter, several factors related to the use of automatic sprinkler systems will be discussed, including their effect on required fire flows, development costs, demand on the fire service, the water purveyor, and the public at large. Finally, some initiatives to encourage sprinkler use will be discussed.

ADVANTAGES OF SPRINKLER SYSTEMS

The following example describes some of the advantages of sprinkler systems. In the example cited, the advantages were translated to dollar savings.

In 1965, in Fresno, Calif., the city fire marshal, the municipal planning department, and other municipal agencies joined in developing an innovative master fire plan. The plan was based on resourceful application of what was known then as the Grading Schedule for Municipal Fire Protection (1956 Schedule with 1964 amendments of the National Board of Fire Underwriters) combined with an integrated system of codes and ordinances. Before developing the master plan, the fire department budget represented approximately 13 percent of a total municipal budget of $14 million. At the time, the fire department was rated as class 1, but the water department was rated only as class 3.

The codes and ordinances adopted as part of the new plan included a dangerous-building ordinance, which gave the fire marshal authority to condemn property deemed unsafe, unsanitary, or dangerous. Owners of condemned property had the choice to sell their buildings to an urban renewal agency or renovate the buildings to code standards, including mandatory automatic sprinkler systems. Under the Federal Urban Renewal Agency Agreement with the city, all new construction required complete sprinkler systems.

In 1955, Fresno covered 21 mi² (13,440 acres), with a population of 115,000 (Grading Schedule for Municipal Fire Protection 1974). The fire department maintained 68 firefighters on duty. In 1977, after full implementation of the master plan, the corporate limits of Fresno covered 58 mi² (37,120 acres), with a population of 184,500. In 1977, 95 percent of all buildings in the urban renewal area were completely fitted with sprinklers. This area alone covered about 40 square blocks of nonresidential property. As a result of the credits then allowed, the fire department budget was reduced to 7.9 percent of the total municipal budget. Only 68 firefighters were still maintained on duty in any given 24-hour period. Fire losses were reduced by 22 percent, and the fire department maintained a class 1 rating.

The savings realized in fire department operation contributed to more efficient distribution of funds under the total municipal budget, and the water department classification rating was changed from 3 to 1. As a result, Fresno was rerated from class 3 to a class 2 municipal rating.

WATER SUPPLY REQUIREMENTS FOR SPRINKLERED PROPERTIES

Sprinklers are designed to control fires but must be applied in conjunction with hose streams to completely extinguish a fire. In large buildings, sprinkler piping is sized to supply water to zones within the building, not to all sprinklers simultaneously. A fire can occasionally overcome the capacity of a sprinkler system. Hose streams are especially important in these situations.

Required fire flow for sprinklered properties consists of the flow required for sprinklers, including a hose-stream allowance or 500 gpm (31.5 L/sec), whichever is greater. The range of the sprinkler requirement varies from 150 to 1,600 gpm (9.5 to 100.9 L/sec). These flows depend on the classification of hazard, whether the system is hydraulically designed or pipe scheduled, the type of materials being stored, and the storage configuration, as well as other factors.

Design Curves

Figure 5-1 shows the design curves used to determine the density required for various hazard classifications. Density is defined as the flow required, in gallons per minute per square foot, to be discharged over a selected area of operation. For example, if the density required is 0.10 gpm/ft² (0.24 m/hr) and it is applied to an area of operation of 1,500 ft² (139.4 m²), the minimum system demand, excluding hose streams, will be 150 gpm (9.5 L/sec). At the opposite extreme, an extra hazard group 2 situation will require 1,600 gpm (100.9 L/sec) (5,800 ft² × 0.275 gpm/ft = 1,595 gpm).

The system demands developed from these curves must be increased, however, to allow for two additional factors. To compensate for friction loss in the piping system, about 10 percent is generally added to the basic system demand. A hose-stream allowance is also included to permit operation of inside and outside hose streams. The total hose-stream allowance and fire flow duration are based on hazard classification in accordance with the schedule in Table 5-1.

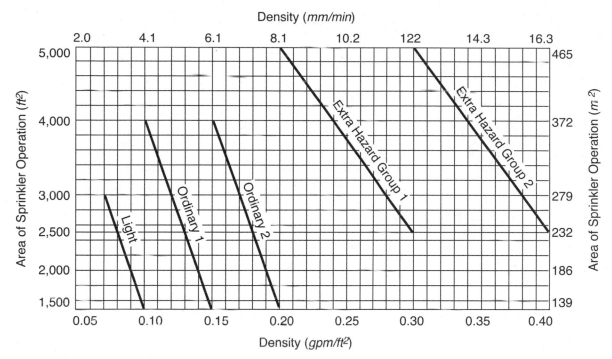

Figure 5-1 Sprinkler system design curves

Reprinted with permission from NFPA 13-2007, Installation of Sprinkler Systems, copyright © 2007, National Fire Protection Association. This reprinted material is not the complete and official position of the NFPA, which is represented solely by the standard in its entirety.

When rack storage up to 20 ft (6.1 m) high is present, the design curves for ceiling sprinklers increase to a maximum of 1,800 gpm (113.5 L/sec) for class 4 unencapsulated commodities. (NFPA 13 covers rack storage of materials.) If the rack storage is 25 ft (7.6 m) high, this basic system demand increases to 3,150 gpm (198.7 L/sec).

Hazard Classifications

It is apparent that while sprinklered buildings are not included when determining the required fire flow for rating purposes, it is necessary to calculate the demand in order to ensure adequate water supplies for sprinklered buildings. Average commercial occupancies, such as retail stores, offices, hotels, and institutional buildings, usually fit the category of ordinary hazard group 1 occupancies. Warehouse and manufacturing occupancies fall in ordinary hazard groups 2 or 3. Occupancies with highly flammable products or processes in large quantities are classed as extra hazard.

The determination of hazard is defined in detail in the NFPA 13 Standard for the Installation of Sprinkler Systems, NFPA 13D Standard for the Installation of Sprinkler Systems in One- and Two-Family Dwellings and Manufactured Homes, and NFPA 13R Installation of Sprinkler Systems in Residential Occupancies up to and Including Four Stories in Height. System demands for potential sprinklered properties must be carefully calculated based on these data. Plans and calculations submitted to the building and/or fire department show the hazard classification.

In residential areas of one- and two-family dwellings, the minimum water supply to be considered for a residential sprinkler system is found in NFPA 13D. Residential

Table 5-1 Hazard Classification Schedule

| Hazard Classification | Hose-Stream Allowance | | Duration |
	gpm	(L/sec)	min
Light	100	(6.3)	30
Ordinary Group 1	250	(15.8)	60–90
Ordinary Group 2	250	(15.8)	60–90
Ordinary Group 3	500	(31.5)	60–120
Extra Hazard Group 1	500	(31.5)	90–120
Extra Hazard Group 2	1,000	(63.1)	120

sprinkler systems designed in accordance with NFPA 13D are required to flow a maximum of only 26 gpm (49 L/min) with a minimum operating pressure of 7 psi at the sprinkler. The flow may be reduced or may be increased depending on the specific listing of the sprinklers used. Pressure requirements will also vary depending on the specific listing of the sprinkler.

TYPES OF SPRINKLERS FOR COMMERCIAL BUILDINGS

The required fire flow for automatic sprinklers is defined in NFPA 13. There are two types of sprinkler systems in general use for commercial buildings—pipe schedule systems and hydraulically calculated systems. A pipe schedule system is a sprinkler system designed by using a pipe-size table with corresponding maximum flow rates from NFPA 13. A hydraulically designed system is a sprinkler system that is designed by using existing or designated water supply pressure and then calculating all water flow rates available for each size of piping in the system. Special sprinkler systems, such as deluge systems, are occasionally installed for special hazards. These special systems are usually hydraulically calculated.

Pipe Schedule Systems

The required fire flow for pipe schedule systems varies from 250 to 1,500 gpm (15.8 to 94.8 L/sec), based on the occupancy of the building. The required fire flow should be available at the base of the sprinkler riser at a pressure equivalent to 15 psi (103.4 kPa) at the highest sprinkler. In these systems, the sprinkler pipe sizes are specified in the appropriate standard, and the piping is arranged like a tree, being 100 percent dead-ended.

Hydraulically Calculated Systems

The required fire flow for hydraulically calculated systems varies from 150 gpm (9.5 L/sec) to several thousand gallons per minute, based on the occupancy and the area of the largest room (up to 5,000 ft² [460 m²]). The required fire flow should be available at the base of the sprinkler riser at sufficient pressure to deliver the required fire flow through the sprinkler piping to the design area, with a minimum pressure at every sprinkler of 7 psi (48.3 kPa). NFPA 13 specifies only a minimum-size pipe and allows for looping the piping. The design of these systems is based on the calculated residual pressure available at the base of the sprinkler riser when delivering the required fire flow at the point of design. If not verified by hydraulic testing, no more than 20 psi (138 kPa) should be assumed. The calculated residual pressure is determined from the available fire flow data in the vicinity. The higher the calculated residual pressure, the smaller the sprinkler piping generally used.

For economic reasons these sprinkler systems are designed using the highest residual pressure at which the required fire flow is available. Unfortunately, this information is usually needed before the building is built and perhaps even before mains are extended to the site. Therefore, it is important that the data used accurately represent the hydraulic conditions expected in the vicinity of the proposed building. For both types of systems, the duration for which the sprinkler system must be designed to operate varies from 30 to 120 minutes, depending on the hazard of the occupancy and the ability to remotely control water flow.

Many formerly adequate sprinkler systems can become inadequate because of a lowering of the normal pressure throughout a community or a general overall intentional lowering of the distribution system pressure.

STANDPIPES

Most local building codes require standpipe installation in any building five stories or 50 ft (15.3 m) high. These standpipes are pressurized vertical pipes in buildings, as opposed to standpipes that are tall ground-level storage tanks. Standpipes are classified as follows: class 1—one 2.5-in. (63.5-mm) fire department outlet on each floor; class 2—one 1.5-in. (38.1-mm) tenant hose station on each floor; and class 3—combined 2.5-in. and 1.5-in. outlets on each floor.

NFPA 14 Installation of Standpipe and Hose Systems is the standard for standpipe systems. The water supply requirements for standpipes, as specified in that standard, are 65 psi (448 kPa) residual pressure at the highest outlet, with 500 gpm (31.5 L/sec) for the first standpipe and 250 gpm (15.8 L/sec) for each additional standpipe, to a maximum of 2,500 gpm (157.7 L/sec). These requirements apply for class 1 and class 3 systems. A class 2 standpipe system requires a total supply of 100 gpm at 65 psi (6.3 L/sec at 448 kPa) residual pressure at the outlets. In fully sprinklered buildings, the water supply must satisfy the more stringent of the two demands (standpipe or sprinklers). Inside hose-stream allowances for hose stations supplied from the sprinkler system (not necessarily class 2 standpipe systems) of 50 gpm (3.2 L/sec) each, to a maximum of 100 gpm (6.3 L/sec), are added to the sprinkler water supply, but no minimum residual pressure is mandated. This inside hose requirement is included in the total hose allowance that was previously discussed.

Private hydrants are required where buildings are so located on the property or are of such size and configuration that a normal hose lay from a public hydrant would not reach all points on the outside of the building. Rules for the number and location of private hydrants are determined by the municipal fire department.

BACKFLOW PREVENTION FOR FIRE SPRINKLER SYSTEMS

Wet-pipe fire sprinkler systems are not intended to supply potable water, therefore these systems may contain materials that are not used in potable water systems. Water in the sprinkler pipeline for an extended time will become stagnant, may degrade in quality, and may not meet drinking water standards. This nonpotable water in the sprinkler system usually requires the need for a backflow prevention assembly. AWWA Manual M14 Recommended Practice for Backflow Prevention and Cross-Connection Control provides guidance on proper backflow prevention practices to assure the drinking water is not contaminated. Many local regulatory authorities (health departments or water suppliers) may have local requirements on the use and installation of backflow prevention assemblies on fire sprinkler systems.

Fire protection systems are engineered to provide proper flow and pressure to the entire sprinkler system. When an existing fire sprinkler system is modified by adding a backflow prevention assembly, or other changes, care must be taken to assure the system will still perform properly. When installing a backflow prevention assembly in a retrofit application, the engineer must be sure to check the hydraulic calculations of the fire sprinkler system to ensure it performs properly.

System Water Pressure for Sprinkler Design

The design of a sprinkler system requires knowledge of the water pressure in the street. However, there is really no such thing as a single, constant water pressure in the street that should be used for design. The pressure in water mains varies over time because of many factors, including

1. Normal daily variations caused by time-of-day water use patterns, tank water level fluctuations, hydraulic transients, valve operation, and cycling of pumps.

2. Long-term system changes caused by water main construction, changes in pressure-regulating valve settings, addition of new pumps, corrosion and scale in piping, and changes in pressure zone boundaries.

3. Long-term variations in water use patterns, including new users and changes in usage for existing users.

4. Short-term emergencies caused by fires, pipe breaks, system components out of service for rehabilitation and repair, power outages, and flows from sprinklers to fight fires.

With all of the sources for variations in pressure, it's clear that there is no single water pressure in the street. Instead, pressure fluctuates over time, and the sprinkler system designer must select a single value as the basis for design from a reasonable worst-case condition.

Owners and designers of sprinkler systems must realize that no pressure can be guaranteed by water utilities 100 percent of the time. There is a finite risk associated with any pressure selected as the basis for sprinkler design.

For example, the pressure in a main may be 55 psi (379 kPa) during a "normal" day with the tanks nearly full and the pumps running; it may be 48 psi (331 kPa) when the tank water level is low and the pumps are off; it may be 42 psi (290 kPa) when the tank water level is low and the water use for lawn watering is high; it may be 32 psi (221 kPa) when there is a major fire a few blocks away; and it may be 0 psi when the main is valved off to repair a pipe break. The state regulatory agency standard for minimum pressure in the water main may be 20 psi (138 kPa). The selection of a pressure to be used for sprinkler system design depends on the level of risk that the owner is willing to accept and the amount of money the owner is willing to spend. This is a subjective decision for which there is no right answer.

Utilities must also be cautious when asked by owners or sprinkler designers about the pressure at a given point in the distribution system. Instead of responding with an answer of "55 psi" (379 kPa), the answer should be that "the pressure was measured on July 23 to be 55 psi during normal pump and tank operating conditions with no extraordinary demands on the system." Utility personnel should avoid giving the impression that it is possible to guarantee a specific pressure under all possible conditions.

Fire sprinkler designers should be aware of the tradeoffs in alternative ways of providing adequate pressure to buildings. On occasion, a sprinkler designer may request, for example, 80 gpm at 65 psi when the distribution system can only provide 55 psi. It is usually much less expensive to modify the sprinkler system to use 55 psi than to modify the distribution system to provide 65 psi.

REFERENCES

Accepted Procedure and Practice in Cross-Connection Control. 1995. American Water Works Association, Pacific Northwest Section.

Air Gaps in Plumbing Systems. ANSI Standard No. 112.1.2-97. New York: American National Standards Institute.

Angele, G.J. *Cross Connection and Backflow Prevention.* 1974. Denver, Colo.: American Water Works Association.

Côté, A.E., and J.L. Linville, eds. 2003. *Fire Protection Handbook.* 19th ed. Quincy, Mass.: National Fire Protection Association.

Double-Check Valve Backflow-Prevention Assembly. AWWA Standard C510. Denver, Colo.: American Water Works Association.

Factory Mutual, Approved Standard. *Backflow Preventers (Reduced Pressure Principle and Double Check Valve Types).* 1977.

Fire Suppression Rating Schedule. 2003. Jersey City, N.J.: Insurance Services Office Inc.

Grading Schedule for Municipal Fire Protection. 1974. New York: Insurance Services Office Inc. (Out of print.)

Installation of Sprinkler Systems. NFPA Standard 13-07. Quincy, Mass.: National Fire Protection Association.

Installation of Sprinkler Systems in One- and Two-Family Dwellings and Manufactured Homes. NFPA Standard 13D-07. Quincy, Mass.: National Fire Protection Association.

Installation of Sprinkler Systems in Residential Occupancies up to and Including Four Stories in Height. NFPA Standard 13R-07. Quincy, Mass.: National Fire Protection Association

Installation of Standpipe and Hose Systems. NFPA Standard 14-07. Quincy, Mass.: National Fire Protection Association.

A Manual to Cross-Connection Control Requirements for Fire Sprinkler Systems. 1979. Patterson, N.Y.: National Fire Sprinkler Association.

Manual of Cross-Connection Control. 1994. 9th ed. Foundation for Cross-Connection Control and Hydraulic Research. Los Angeles, Calif.: University of Southern California.

Performance Requirements for Reduced Pressure Principle Backflow Preventers. ASSE Standard 1013-93. Bay Village, Ohio: American Society of Sanitary Engineers.

Recommended Backflow Prevention and Cross-Connection Control. AWWA Manual M14. Denver, Colo.: American Water Works Association.

Reduced-Pressure Principle Backflow-Prevention Assembly. AWWA Standard C511. Denver, Colo.: American Water Works Association.

Appendix A

Agencies Involved in Fire Protection

The functions and policies of the various agencies that are involved in fire protection are described in the following paragraphs.* Those agencies that have contributed reference materials, guidance, and support to the development of this manual are described.

INSURANCE SERVICES OFFICE INC.

The Insurance Services Office Inc. (ISO), is a corporation that provides services for participating companies. It provides statistical, actuarial, and survey information for numerous affiliated insurance companies and more than a dozen different lines of insurance. Included in these lines are liability, automobile, boiler and machinery, home owner's, farm, and commercial fire insurance.

The various functions of the ISO were, before the early 1970s, performed by different organizations in different states. Currently, the ISO gathers data used to establish loss costs that may be used by member insurance companies to determine rates for their fire protection policies for both residential and commercial properties. One basic philosophy of the ISO, which is gaining wide acceptance, is the belief that building developers and owners can do far more to protect and limit loss caused by fire than can any community fire protection system.

The ISO addresses the various facets of public protection to determine what might occur at a specific location when all design systems in the building fail and a fire occurs. It is concerned with the potential for a public fire protection system to avoid a total loss and to significantly minimize loss by manual fire suppression activities. The ISO is not actively involved in influencing local government officials to establish policies or procedures concerning local fire protection systems.

---

* The information in this appendix is adapted from the 19th edition of the *Fire Protection Handbook,* published by the National Fire Protection Association, Quincy, Mass.

The basic document by which the ISO evaluates the potential loss-limiting opportunities of the public fire suppression system is the *Fire Suppression Rating Schedule.*

INSURANCE ORGANIZATIONS WITH FIRE PROTECTION INTERESTS

In addition to the ISO, a number of other insurance organizations are active in fire protection. They include the American Insurance Association, the Alliance of American Insurers, and a significant number of other agencies with relatively narrow areas of interest.

NATIONAL FIRE PROTECTION ASSOCIATION

The National Fire Protection Association (NFPA) is a scientific and educational organization primarily involved in studying and disseminating information concerning the causes, prevention, and control of losses caused by fire. Chapter 4 of the current edition of the *Fire Protection Handbook* (published by NFPA, 19th ed.) contains background material dealing with the role of the water distribution system in fire protection. This information is of particular interest and was of considerable assistance to the AWWA committee in developing this manual.

NATIONAL FIRE SERVICE ASSOCIATIONS

A number of fire service associations serve the basic needs of people involved in fire protection. Some of these are the Fire Marshals Association of North America, the Fire Service Section of the National Fire Protection Association, the International Association of Arson Investigators, the International Association of Black Professional Firefighters, the International Association of Fire Chiefs, the International Association of Firefighters, the International Fire Service Training Association, the International Municipal Signal Association, the International Society of Fire Service Instructors, the Joint Council of Fire Service Organizations, and the National Volunteer Fire Council.

FIRE RESEARCH LABORATORIES

Many laboratories throughout the United States perform varying degrees of fire-related research. One prominent laboratory is the Underwriters Laboratories, a nonprofit organization. Its objective is to promote public safety through scientific investigations, experiments, tests, and the publication of standards and specifications for materials, devices, and other items that tend to reduce or prevent bodily injury, loss of life, or loss of property from fire-related hazards.

Another prominent research laboratory is the Factory FM Global, whose objective is to research the nature of fire and to develop standards for establishing practices to minimize loss due to fires.

NATIONAL FIRE SPRINKLER ASSOCIATION

The National Fire Sprinkler Association (NFSA) is a trade association comprised of manufacturers and installing contractors of automatic fire sprinkler systems and related equipment. Organized in 1914, it has continuously served the fire sprinkler industry, operating under the name of National Automatic Sprinkler and Fire Control Association until 1983, at which time it became NFSA.

The association represents the fire sprinkler industry in a variety of national programs, including labor relations, market development, education and training, and public relations. These functions are carried out through close liaison with organizations that write model building codes and standards, the insurance underwriting industry, and other national organizations, including the National Fire Protection Association, Construction Specifications Institute, American Water Works Association, Society of Fire Protection Engineers, National Institute for Standards and Technology, and others.

The association is governed by a board of directors representing manufacturers, contractors, and suppliers. Board policies are implemented by a professional staff headquartered in Patterson, N.Y.

AMERICAN FIRE SPRINKLER ASSOCIATION

The American Fire Sprinkler Association (AFSA) is a nonprofit, international association representing open shop fire sprinkler contractors and dedicated to the educational advancement of its members and promotion of the use of automatic fire sprinkler systems.

INTERNATIONAL CODE COUNCIL

The International Code Council, a membership association dedicated to building safety and fire prevention, develops the codes used to construct residential and commercial buildings, including homes and schools. Most US cities, counties, and states that adopt codes choose the International Codes developed by the International Code Council.

INSURANCE BUREAU OF CANADA

The Insurance Bureau of Canada acts as the insurance trade association representing, in general, the interests of the insurance industry in Canada.

INSURERS' ADVISORY ORGANIZATION

The Insurers' Advisory Organization (IAO), based in Toronto, Ont., is somewhat similar to the US-based ISO. It is an independent agency that provides advisory services to its membership in the insurance industry.

One of the IAO sections is the Fire Underwriters Survey (FUS), which is sponsored by the Insurance Bureau of Canada. A main function of FUS is to survey the fire protection condition of Canadian municipalities and to provide data and advisory services for such municipalities and insurance companies. The 1974 edition of the *Grading Schedule for Municipal Fire Protection* was authored and used by ISO until 1980. FUS also published a booklet called *Water Supply for Public Fire Protection: A Guide of Recommended Practice*.

Standards and manuals issued by the NFPA are available in Canada through the Canadian Association of Fire Chiefs.

NATIONAL RESEARCH COUNCIL

The National Research Council, a federally sponsored agency located in Ottawa, Ont., is active in research and studies on fire protection.

UNDERWRITERS LABORATORIES OF CANADA

The Underwriters Laboratories of Canada (ULC) is active in fire-related research and testing. NFPA standards and ULC listings of equipment and materials are often part of or serve as complementary documents to the National Building Code of Canada, the Provincial Building and Fire Codes, and the Municipal Building Codes and Bylaws.

Appendix **B**

Water Supply and Fire Insurance Ratings

INSURANCE RATINGS

Adequate water supply and fire hydrants can make it possible to more effectively fight fires. The adequacy of the water supply is one of the factors used by insurance companies in developing insurance rates for individual properties.

This appendix will focus on what are called public protection systems—public water supply systems that are called on to provide from 500 to 3,500 gpm (31.5 to 221 L/sec) for fire protection. In this section, *cities* refers to a city, town, or other political subdivision. Some large cities have their grading based on actual loss history.

Water systems are rated (or graded) by insurance organizations or independent rating bureaus. The most widely used grading schedule is the *Fire Suppression Rating Schedule* (2003) developed by the Insurance Services Office Inc. (ISO). The ISO rating system has evolved from a system initially developed by the National Board of Fire Underwriters in 1889. The 2003 *Fire Suppression Rating Schedule* superseded the *Fire Protection Grading Schedule,* which involved a considerably more detailed process and placed greater emphasis on reliability of water system components.

The *Fire Suppression Rating Schedule* is used by ISO to objectively review and correlate those features of public fire protection that have a significant effect on minimizing fire damages at commercial properties. Water utilities desiring copies for reference purposes may obtain them at

Insurance Services Office Inc.
545 Washington Blvd.
Jersey City, NJ 07310-1686

The *Fire Suppression Rating Schedule* assigns a city classification on a scale of 1 to 10 based on the fire department, fire alarm system, and water supply. The highest possible rating of 1:10 represents a city with no recognized fire department. These ratings are calculated using a formula in the *Fire Suppression Rating Schedule* and are rounded to the nearest integer. This integer value is referred to as the Public Protection Classification (PPC™) of the city. Most cities in the United States have been assigned a Public Protection Classification. The actual insurance rates for specific properties are set by individual insurance companies based on the Public Protection Classification and the market.

Water System Rating

The ISO system for rating water supply is described in detail in the *Fire Suppression Rating Schedule* and is summarized in Figure B-1. Of the 100 points assigned in rating a city, 40 depend on the water supply system. Two points are assigned based on hydrant size, type, and installation, and three points are assigned based on inspection and condition of hydrants.

The remaining 35 points are simply based on a comparison of the needed fire flow with the *capability of the water system* at test locations. Methods to determine needed fire flow are described in chapter 1 of this manual, and additional detail is provided in the *Fire Suppression Rating Schedule*.

The capability of the fire system is based on the lowest of three factors: (1) supply works capacity, (2) main capacity, and (3) hydrant distribution. The supply works capacity is the excess capacity of the water treatment and pumping system (in gpm) above that required to meet maximum day demand. Storage can be used to supplement this capacity. The main capacity is the discharge (in gpm) that can be delivered to the test location at 20 psi (138 kPa). Hydrant distribution refers to the amount of water that can be delivered through hydrants within given distances of the test location.

If the water supply capability equals or exceeds the needed fire flow, the full credit is given at that test location. If not, fewer than 35 credits are given at that location.

System Testing

The main capacity of the distribution system is based on the results of actual fire hydrant flow tests conducted in accordance with ISO test procedures that are summarized in AWWA Manual M17, *Installation, Field Testing, and Maintenance of Fire Hydrants*. Locations for flow tests are selected to give a wide range of occupancies, construction types, elevations, building spacings, and needed fire flows. Tests include the buildings with the five highest fire flows. Unless a city is primarily residential, no more than 25 percent of the flow tests are conducted in residential neighborhoods.

The number of test locations depends on the number of hydrants in the city and ranges from as few as two to three for cities with fewer than 10 hydrants to as many as 25 to 75 for cities with more than 8,000 hydrants. The locations are selected by ISO and are generally conducted during a weekday unless a different time is specifically requested in congested areas.

Cities are notified in advance of fire flow testing. It is good practice to notify customers in the vicinity of flow tests especially if problems with discolored water are anticipated.

Improving Public Protection Rating

The rating system is based on the condition of the system at a single point in time and does not take into account components that are in the planning stage or are out of service for maintenance. It is best not to evaluate a system when a major system component is out of service.

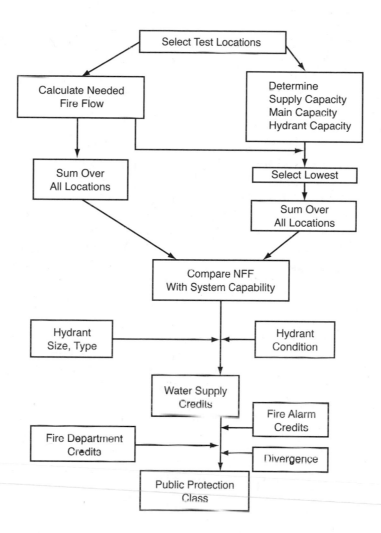

Figure B-1 Water supply evaluation

Usually, main capacity is evaluated based on an actual fire flow test. However, benchmark results of tests are used to compare with the results of computer modeling studies, when they are available. If the benchmark test results are within a certain tolerance of the computer modeling predictions, these predictions may be used in lieu of additional flow test results.

Because water system capability is based on the worst case of the treatment, main, and hydrant capacity, a city with a less than perfect rating should review the weakest component if it desires to improve its rating. For example, if the needed fire flow at a location is 1,500 gpm (95 L/sec) and the supply, main, and hydrant capacities are 2,500, 1,200, and 2,000 gpm (158, 76, and 126 L/sec), respectively, the city should evaluate main improvements if it wants to improve its rating. If all three flows were to exceed needed fire flow, then water system improvements may not affect insurance ratings.

Even if the water system is improved, it may not affect insurance ratings if it does not move the city to a lower or better public protection class. Another rating factor is that of *divergence*, meaning that the water and fire departments should be of similar quality. Cities can lose points if the water supply is substantially better than the fire department, or vice versa.

Regional or private water utilities must work with the cities they serve to establish standards and cost allocation for improvements for fire protection purposes. Costs for components such as fire hydrants or extra storage for fire protection can be paid out of utility revenue, general city revenue, or special sources (e.g., developer contributions).

Interaction with ISO

At the beginning of a water system survey, ISO representatives meet with city officials to review maps of the city, including water distribution maps. In addition to observing flow tests, they visit major pumping and treatment and storage facilities.

ISO sends a report to the chief executive officer of a community giving the new protection class rating and a summary grading sheet, as well as classification details and improvement statements. A single public protection class is assigned to an entire city unless a significant portion (more than 15 percent) does not have fire hydrants. In that case, that portion of the city would have a different rating.

A city may request a reevaluation during the interim period if it has made major improvements to its system.

REFERENCES

Installation, Field Testing, and Maintenance of Fire Hydrants. AWWA Manual M17. Denver, Colo.: American Water Works Association.

Fire Suppression Rating Schedule. 2003. Jersey City, N.J.: Insurance Services Office Inc.

Index

This page intentionally blank.

AWWA Manuals

M1, *Principles of Water Rates, Fees, and Charges,* Fifth Edition, 2000, #30001PA

M2, *Instrumentation and Control,* Third Edition, 2001, #30002PA

M3, *Safety Practices for Water Utilities,* Sixth Edition, 2002, #30003PA

M4, *Water Fluoridation Principles and Practices,* Fifth Edition, 2004, #30004PA

M5, *Water Utility Management Practices,* Second Edition, 2006, #30005PA

M6, *Water Meters—Selection, Installation, Testing, and Maintenance,* Second Edition, 1999, #30006PA

M7, *Problem Organisms in Water: Identification and Treatment,* Third Edition, 2004, #30007PA

M9, *Concrete Pressure Pipe,* Second Edition, 1995, #30009PA

M11, *Steel Pipe—A Guide for Design and Installation,* Fifth Edition, 2004, #30011PA

M12, *Simplified Procedures for Water Examination,* Fifth Edition, 2002, #30012PA

M14, *Recommended Practice for Backflow Prevention and Cross-Connection Control,* Third Edition, 2003, #30014PA

M17, *Installation, Field Testing, and Maintenance of Fire Hydrants,* Fourth Edition, 2006, #30017PA

M19, *Emergency Planning for Water Utility Management,* Fourth Edition, 2001, #30019PA

M20, *Water Chlorination/Chloramination Practices and Principles,* Second Edition, 2006, #30020PA

M21, *Groundwater,* Third Edition, 2003, #30021PA

M22, *Sizing Water Service Lines and Meters,* Second Edition, 2004, #30022PA

M23, *PVC Pipe—Design and Installation,* Second Edition, 2003, #30023PA

M25, *Flexible-Membrane Covers and Linings for Potable-Water Reservoirs,* Third Edition, 2000, #30025PA

M27, *External Corrosion—Introduction to Chemistry and Control,* Second Edition, 2004, #30027PA

M28, *Rehabilitation of Water Mains,* Second Edition, 2001, #30028PA

M29, *Fundamentals of Water Utility Capital Financing,* Third Edition, 2008, #30029PA

M30, *Precoat Filtration,* Second Edition, 1995, #30030PA

M31, *Distribution System Requirements for Fire Protection,* Third Edition, 1998, #30031PA

M32, *Distribution Network Analysis for Water Utilities,* Second Edition, 2005, #30032PA

M33, *Flowmeters in Water Supply,* Second Edition, 2006, #30033PA

M36, *Water Audits and Leak Detection,* Second Edition, 1999, #30036PA

M37, *Operational Control of Coagulation and Filtration Processes,* Second Edition, 2000, #30037PA

M38, *Electrodialysis and Electrodialysis Reversal,* First Edition, 1995, #30038PA

M41, *Ductile-Iron Pipe and Fittings,* Second Edition, 2003, #30041PA

M42, *Steel Water-Storage Tanks,* First Edition, 1998, #30042PA

M44, *Distribution Valves: Selection, Installation, Field Testing, and Maintenance,* Second Edition, 2006, #30044PA

M45, *Fiberglass Pipe Design,* Second Edition, 2005, #30045PA

M46, *Reverse Osmosis and Nanofiltration,* Second Edition, 2007, #30046PA

M47, *Construction Contract Administration,* First Edition, 1996, #30047PA

M48, *Waterborne Pathogens,* Second Edition, 2006, #30048PA

M49, *Butterfly Valves: Torque, Head Loss, and Cavitation Analysis,* First Edition, 2001, #30049PA

M50, *Water Resources Planning,* Second Edition, 2007, #30050PA

To order any of these manuals or other AWWA publications, cal lthe Bookstore toll-free at 1.800.926.7337.

M51, *Air-Release, Air/Vacuum, and Combination Air Valves*, First Edition, 2001, #30051PA

M52, *Water Conservation Programs—A Planning Manual*, First Edition, 2006, #30052PA

M53, *Microfiltration and Ultrafiltration Membranes for Drinking Water*, First Edition, 2005, #30053PA

M54, *Developing Rates for Small Systems*, First Edition, 2004, #30054PA

M55, *PE Pipe—Design and Installation*, First Edition, 2006, #30055PA

M56, *Fundamentals and Control of Nitrification in Chloraminated Drinking Water Distribution Systems*, First Edition, 2006, #30056PA